NASA, The First 25 Years 1958–1983

A Resource for Teachers

A curriculum project

NASA

National Aeronautics and
Space Administration
Washington, D.C.
1983

For sale by the Superintendent of Documents,
Government Printing Office, Washington, DC 20402.

Table of Contents

Introduction

In 1958 a unique Federal agency was established with a mandate from the Congress to "plan, direct, and conduct aeronautical and space activities." That simply stated charge to the National Aeronautics and Space Administration began 25 years of aeronautical and space programs and projects that brought dreams to reality, made engineering ideas into technological accomplishments, developed practical applications of space research, presented ever-new frontiers of science, and at last effected regular operational service of the Shuttle. These 25 years of space ventures and discoveries and the excitement they engendered were shared by millions. For the new agency had another mandate: to "provide for the widest practicable and appropriate dissemination of information concerning its activities and the results thereof."

Thus, in 1960, NASA established an office to serve the educational community. Our staff was small, the programs few. Aerospace specialists travelled to schools with the Spacemobile's assembly program and assisted aerospace education summer workshops for teachers, there were NASA awards at the International Science and Engineering Fair (ISEF), and the first materials for the classroom teacher, K-12, were produced.

As NASA's projects were developed, conducted, and completed, our educational office and programs grew apace. The staff now encompasses a branch in the Public Affairs Division at NASA Headquarters in Washington, D.C. and educational programs officers in seven NASA Centers that serve specific geographic regions. In 1983, the Spacemobile is but one part of the Aerospace Education Services Project (AESP) and now provides lecturers who work in classrooms as well as assembly halls; the simple experiments and scale models of space hardware used in their lectures have changed through the years with each new NASA activity. There is also an Aeronauticsmobile that visits schools to discuss NASA research and development in that field. Two years ago the AESP introduced two new programs: the Urban Community Enrichment Program (UCEP) to stimulate learning at the middle school level in large cities, and CLASS (College Lecturers on Aeronautics and Space Sciences), to bring a better understanding of the agency's research and development activities to college and university engineering and science students.

Our ISEF program soon added NASA awards for affiliated state and regional science fairs. Next, there were Youth Science Congresses followed by the Skylab Student Project which enabled 19 high school students to fly experiments on the spacecraft, and a Viking Student Project to select an emblem for the Viking Lander spacecraft. Since 1980 the Shuttle Student Involvement Project has given high school students an opportunity to develop experiments for Shuttle flights. At several NASA Centers there are academic year and summer programs for selected local students, and career information related to NASA and aerospace industry has been made available.

Through the years we have also responded to the community, making both educational staff and AESP specialists available for civic clubs and professional organizations. Audiovisual consultants assist in the programming of materials for radio and television, and we provide support to programs operated by planetariums, museums, and science centers. On invitation from individual communities, our "Community Involvement Programs" involve the entire spectrum of the local populace in an aerospace event.

We have developed a strong audiovisual program over the years, which now includes the production of films, with teaching guides, for specific classroom use; full-length planetarium programs; and curriculum materials on videotape. We attempt to stay in the forefront in the use of educational technology—last year, the laser disc was added to the lecture programs and aerospace activities for microcomputers have been developed for teacher workshops.

Another special program developed for teachers is the Lunar Sample Educational Packet, a lunar and planetary sciences teaching aid using samples of lunar material encapsulated in a clear plastic disc. And several Visitor Information Centers at NASA installations have established resource rooms for educators.

From the outset, our educational publications have covered a variety of subjects—the broad scope of the agency's programs, the aims of individual projects, the specific results of others—and from the initial booklets and folders has come a wide range of materials. There are curriculum supplements, bibliographies, single resource units, reprints from professional journals, and explanatory briefs. There are teacher's guides for specific publications as well as classroom films, and the *NASA Report to Educators,* a quarterly newsletter.

Often, publications have been suggested by teachers searching for pertinent up-to-date information with which to enrich their programs. *NASA, The First 25 Years* is a response to such requests. A record book of aerospace facts, it differs from earlier publications in both subject and form. Today's high school students were born after the first satellites discovered the Van Allen radiation belts and revolutionized communications and meteorology, and after men had orbited Earth; today's elementary students did not watch the lunar landings. For their teachers, this is a summary of the important dates, projects, goals, and achievements that are history for their charges. With an introduction outlining the United States' research in air and space technology during the first half of the century, it is intended as a ready reference, an historical overview of NASA's first 25 years, the growth and scope.

The Shuttle has fostered a new enthusiasm and interest among school students who have grown up with daily satellite weather maps and with the Moon and planets as places to visit. We hope this compilation of outstanding events since 1958 not only celebrates NASA's 25th anniversary, but also contributes to today's classrooms a sense of the unparalleled wonder and excitement that accompanied each succeeding event. It may even serve as a springboard for the imaginations of youngsters whose ideas will become reality in their future, the next 25 years.

NASA's Technical Monitor and editor for this project was Muriel M. Thorne, Educational Programs Officer, under the general direction of William D. Nixon, Chief of Education Services, NASA.

National Aeronautics and
Space Administration
Washington, D.C.
May 1983

Preface

The United States has been in space for 25 years. During these years we have seen many "firsts," including the first American satellite in orbit, the first Americans in space, the first humans to set foot on the Moon. We have pushed the state of the art in aeronautical research and probed the secrets of planets in our solar system. To remember or to recall each of the many accomplishments would be difficult. We have compiled on the 25th anniversary this historical resource book to give teachers easy access to NASA activities since the agency was founded in 1958.

We are grateful to the many scientists and engineers at NASA Headquarters who have supported the research, writing, and editing of this book. Particular thanks go to Muriel Thorne of the Education Services Branch. We are also indebted to the Public Affairs Officers for the several program offices: Kenneth C. Atchison, Aeronautics and Space Technology; David W. Garrett and James F. Kukowski, Space Flight; Debra J. Rahn, International Affairs; and Charles R. Redmond, Space Science and Applications.

Jane D'Alelio
Jane Tully
Wendy Cortesi

Washington, D.C.
May 1983

Foreword

Because it is impossible to describe the 25 years of NASA's research and missions in detail, this book is designed to provide a reference base from which teachers can develop classroom concepts and activities.

It begins with a prologue, a brief history of the National Advisory Committee for Aeronautics, NASA's predecessor. Chapter I introduces NASA—the agency, its physical plant, and its mission. Succeeding chapters are devoted to major NASA programs, in alphabetical order; within the chapters projects are listed chronologically. Each chapter concludes with ideas for the classroom and space for notes and new information the user may wish to add. The epilogue offers some perspectives on these first 25 years and a glimpse of the future. Appendices include a record of NASA launches and a list of the NASA educational services offices.

For detailed research the teacher should: Request a current catalogue of publications with a price list from the Superintendent of Documents, Government Printing Office, Washington, DC 20402; obtain a copy of NASA's annotated *Aerospace Bibliography, Seventh Edition* from the GPO (Stock No. 033-000-00861-9, $6.00); and, of course, consult the indispensable *Readers' Guide to Periodical Literature.*

Prologue

National Advisory Committee for Aeronautics

Nineteen hundred. The first Age of Flight was 117 years old and balloons were both transportation and sport. Jules Verne that year would see many of his fictional ideas as technical fact at the Paris World's Fair. His younger contemporary, H. G. Wells, was a recognized author of space and time fiction. Orville and Wilbur Wright, two brothers who owned a bicycle shop in Dayton, Ohio, were preparing to test their newly-invented glider on the North Carolina Outer Banks. And 17-year-old Robert Goddard held fast a vision of spaceflight that had come to him when he climbed a cherry tree.

Three years later Orville and Wilbur Wright changed the world.

On December 17, 1903, at 10:35 a.m. near Kitty Hawk, North Carolina, in a 27-mile-an-hour wind from the north, Orville Wright made the first powered flight. 12 seconds. 120 feet. The machine with the homemade 12-hp engine rose from the ground and

First Wind Tunnel. Dedicated in 1920, this wind tunnel was actually put into operation in 1918 at the Langley Memorial Aeronautical Laboratory.

landed at a point as high as the one from which it started. Quietly, unheralded, in an isolated spot the two brothers fulfilled an age-old dream and the age of heavier-than-air flight was reality.

After the Wrights' invention became known, progress in aeronautics was rapid and continual with new flight records established regularly. During the next decade: The Wrights sold their Military Flyer to the U.S. Army; in 1906 Europe saw its first powered flight; the English Channel was crossed; international meets took place on both sides of the Atlantic; the Wright Factory in Dayton produced aircraft manufactured to a standard design pattern. There was radio from plane to ground, a landing on the deck of a ship, and the first practical seaplane. The first U.S. transcontinental flight from Long Island to California took 49 days, including 82 hours, 2 minutes flying time, and 70 landings. 1912 was the year for the first parachute jump from an airplane and when the aircraft speed record was 108.17 mph.

The first regularly scheduled airline in America began operation in 1914. Passenger and cargo potential were being recognized. Factories and flying schools were established. When World War I came, every major nation had aeronautical research facilities—except the United States.

NACA

It was this lack of a government laboratory devoted to the science of flight that prompted the creation of

Orville Wright at the controls of the Wright Flyer as the first flight was made on December 17, 1903. Wilbur Wright, running alongside one wing tip, was able to keep up with it.

the National Advisory Committee for Aeronautics, or NACA. It was founded in 1915, just before the United States' entry into the War, to bring competence to the backwardness of American aviation.

For 43 years the NACA excelled in carrying out its chartered mandate ". . . to supervise and direct the scientific study of the problems of flight, with a view of their practical solution."

The Committee first surveyed the current stage of development of aircraft and the research needs of aeronautics, then set about building the scientific staff and unique research facilities required.

Facilities

In June 1920, the first laboratory, the Langley Memorial Aeronautical Laboratory in Hampton, Virginia, was dedicated; aerodynamics became the major research effort and wind-tunnels the chief tool. Within ten years the results were impressive and recognition worldwide: The up-to-date wind tunnels were hailed as far-seeing; the NACA cowling (1928) for air-cooled radial engines, a streamlined shape that increased aircraft speed, led to the low-wing multi-engine air transports and bombers of the 1930s; systematic studies of aerodynamic drag reduction improved design practices, including the advantages of retractable landing wheels over fixed, exposed landing gear.

A second research center, the Ames Aeronautical Laboratory, was constructed near San Francisco in 1939 with a wind tunnel that dwarfed its predecessor at Langley. A third facility, which was later named the Lewis Flight Propulsion Laboratory, was built in Cleveland in 1940 to perform basic research, develop and test aircraft engines, and study fuels. Research on the jet engine began there in 1943.

The second World War focused aeronautical research on combat aircraft and NACA work on aerodynamics and structural research resulted in extremely effective fighter planes.

Postwar research at higher speeds led to high-altutude drop-test models to gather flight data; then, to using rockets to launch models to transonic (speeds from just below to just above the speed of sound) and supersonic speeds. Langley acquired a surplus naval station on Wallops Island, Virginia, and called it the Pilotless Aircraft Research Division. Next, a High-Speed Flight Research Station was established at Muroc (later Edwards), California, for a series of special research aircraft.

Research

During the years of aeronautical progress, a lone figure had been investigating rockets as "A Method of Reaching Extreme Altitudes." Dr. Robert H. Goddard published such a titled paper in 1919. Seven years later he launched the first liquid propellant

First meeting of the National Advisory Committee for Aeronautics, April 23, 1915. Seated, left to right: Dr. William F. Durand, Stanford University; Dr. S. W. Stratton, Director, National Bureau of Standards; Brig. Gen. George P. Scriven, Chief Signal Officer, War Department; Dr. Charles F. Marvin, Chief, U.S. Weather Bureau; Dr. Michael I. Pupin, Columbia University. Standing, left to right: Holden C. Richardson, Naval Constructor; Dr. John F. Hayford, Northwestern University; Capt. Mark L. Bristol, Director of Naval Aeronautics; Lt. Col. Samuel Reber, Signal Corps, in Charge of Aviation Section. Also present were Dr. Joseph S. Ames, Johns Hopkins University and the Hon. B. R. Newton, Assistant Secretary of the Treasury (above).

Dr. Richard T. Whitcomb with a research model of a supersonic transport. The indentation in the fuselage incorporated the Area Rule design concept he discovered (right).

rocket. For the next two decades, he conducted research, built and flew rockets, provided a mathematical analysis for multistage rockets, and amassed more than 150 patents by the time of his death in 1945.

By the end of World War II NACA's research had led to rocket propulsion and air and space flight had met. The X-series of rocket research aircraft began in 1944. The X-1 was built specifically to investigate the

transonic region and to break the sound barrier. On October 14, 1947, Air Force Capt. Charles E. Yeager piloted the X-1 through the speed of sound for the first time. Beginning with the X-1's historic flight, the Research Airplane Program successfully provided a series of flight vehicles that explored areas of performance and effects of designs in the transonic and supersonic regions for more than 20 years.

At Langley a transonic wind tunnel was created in 1950, a tool that researcher Richard T. Whitcomb used in discovering the "area rule" (the cross-section areas of an aircraft should not alter too rapidly from the front to back of a plane). A genuine breakthrough in airplane design, its immediate application allowed military aircraft to break the sound barrier in level flight.

Most famous of the X research planes was the X-15. An idea in 1952, it achieved its designed altitude and speed objectives in 1968, thus spanning the transition from aeronautical research to the new Space Age.

The X-1, first aircraft to fly faster than the speed of sound in level flight.

International Geophysical Year

The International Geophysical Year (IGY) was observed from July 1957 to December 1958. Its scientific program included a proposal to launch satellites that would measure Earth from space. Russia orbited Sputnik on October 4, 1957, and the United States, Explorer 1, four months later. The first U.S. satellite was launched by the Army. Named Explorer—its mission was to explore the unknown—the satellite fulfilled America's commitment to the IGY. And its small package of instruments produced the first major discovery of the Space Age, the Van Allen radiation belts surrounding Earth.

January 31, 1958. The Jupiter-C rocket on the launch pad at Cape Canaveral prior to launching Explorer 1.

NACA Becomes NASA

The success of Sputnik spurred the creation of a new agency to develop a national space program, which President Eisenhower wanted to emphasize the peaceful uses of research and development. Three agencies vied for leadership: the Atomic Energy Commission, the Department of Defense, and NACA.

The NACA proposal combined aeronautic and space research with a solid scientific base. The Committee also could offer experience in working closely with the military as well

Flight Research Team. Pilot and scientist worked together in early aeronautical research tests at the Langley Laboratory (right).

Dr. Robert H. Goddard standing beside the first liquid-propellant rocket which flew at Auburn, Massachusetts, March 16, 1926 (below).

as providing research for civil applications. By April 1958, the administration's position and the NACA proposal had been combined into a bill for creating a national aeronautics and space agency. On July 29, President Eisenhower signed into law the National Aeronautics and Space Act of 1958.

NACA was a service agency. By discharging its primary responsibility—scientific laboratory research in aeronautics—it both served the needs of all Government departments and coordinated aeronautical research in the U.S. Through membership on committees and subcommittees, it linked government agencies concerned with flight, the aviation and allied industries, and education and scientific institutions; through sponsored research, symposia, and technical conferences and reports, it distributed research information.

The new agency had NACA as its nucleus. The NACA staff, facilities, programs and responsibilities were transferred to NASA. Its tradition of excellence also was a legacy to the new organization when NACA ceased to exist on September 30, 1958.

For the Classroom

1. Have your students prepare a timeline of human interest in flight, of a history of flight, of powered flight.
2. Research topics:
   - The precursors of the Wright Brothers
   - The X-series of research aircraft
   - Women in aviation
   - Flight between 1903 and 1918
   - Flight as sport and as transportation
   - The history of rocketry
3. For book reports, have your students select biographies of aviation and space pioneers.

NATIONAL AERONAUTICS AND SPACE ADMINISTRATION
U.S.A.

I

National Aeronautics and Space Administration

On October 1, 1958, the National Aeronautics and Space Administration came into being ". . . devoted to peaceful purposes for the benefit of all mankind."

NACA Headquarters staff in Washington and their colleagues in the three laboratories and two flight stations became the foundation of the new organization. Later, two Army programs were transferred: the Development Operations Division of the Army Ballistic Missile Agency at Redstone Arsenal in Huntsville, Alabama, and the Jet Propulsion Laboratory in Pasadena, California.

Organization

From the start, NASA was a network of centers and facilities across the United States with its headquarters in Washington, D.C.

Headquarters

The NASA Headquarters offices manage the spaceflight centers, research centers, and other installations. The staff has responsibility for determining projects and programs; establishing management policies, procedures, and performance criteria and review; and analysis of all phases of the aerospace program.

As with any vital, growing, active organization there have been many reorganizations during NASA's 25 years. Programs have been initiated, conducted, concluded. Directions have changed and management has adapted to each succeeding realignment. But the overall mission has remained—aerospace research and development for the benefit of all—which can be introduced through the program offices.

The Office of Aeronautics and Space Technology has two primary responsibilities: In aeronautics, to develop the technology needed to assure safer, more efficient, economical, and environmentally acceptable air transportation systems; in space research and technology, to provide a technology base to support current and future space activities, to coordinate the agency's total program of supporting research and technology related to carrying out specific flight missions to insure an integrated and balanced agency research program, and to coordinate NASA's support of other federal agencies in energy research and development.

The Office of Space Science and Applications is responsible for research and development activities in Earth resources; meteorology; communications; life sciences; and, by using a variety of flight systems and ground-based observations, to increase knowledge of the universe.

The Dolley Madison House, named for the President's wife who lived there for several years, was the site of NASA's first Headquarters.

The Office of Space Flight is responsible for the research, development, and operations of spaceflight programs, including the Space Shuttle.

The Office of Tracking and Data Systems is responsible for the development, implementation, and operation of tracking, data acquisition, command, communications, data processing facilities, and systems and services required to support NASA flight missions.

Facilities

In addition to Headquarters there are now ten NASA field installations and a contract-operated laboratory. Three were specifically sited and constructed for NASA. A broad range of research and development activities is conducted in the installations by government-employed scientists, engineers, and technicians who also manage contracts with universities and industries. Its many laboratories—subsonic, transonic, supersonic wind tunnels; propulsion test facilities; elaborate computer systems; flight simulators; flight test capabilities—have rightly been called a national resource.

Ames Research Center (ARC)

When NACA's Ames Laboratories in Mountain View, California, became

NASA's first leaders. Hugh L. Dryden is presented his commission as deputy administrator by President Eisenhower with T. Keith Glennan, administrator, looking on.

NASA's Ames Research Center, it continued to focus on basic and applied research in the aeronautical, physical, space, and life sciences. Its programs include short and vertical takeoff and landing (STOL and VTOL) technology, operation of NASA's Kuiper airborne observatory, and Pioneer, its first space project.

Dryden Flight Research Facility (DFRF)

NACA's High Speed Flight Station at Edwards, California, was first renamed Flight Research Center and later, for Hugh L. Dryden, long-time director of NACA and NASA's first deputy administrator. Its functions have always been related to aeronautical programs—the X series of research aircraft, lifting bodies, and now the design of remotely-piloted vehicles. In 1978 it was the site of the Shuttle's Approach and Landing Tests and served as landing site for five of the first six Shuttle orbital flights.

Goddard Space Flight Center (GSFC)

Named for America's rocket pioneer, the Goddard Space Flight Center at Greenbelt, Maryland, was the first facility built for NASA. Its main responsibility has concerned the design, development, and construction of Earth-orbiting scientific and application satellites and their tracking and data analysis. Its first program, Explorer 6, was followed by other Explorers, several observatory spacecraft, and meteorological and Earth resource satellites. It manages the sounding rocket program, and the Tracking and Data Relay Satellite System (TDRSS) is under its aegis. GSFC also operates the National Space Science Data Center for storing and distributing information gained from NASA Earth-orbital and deep space missions, and, in New York City, the Goddard Institute for Space Studies.

Jet Propulsion Laboratory (JPL)

NASA-owned and contract-operated by the California Institute of Technology in Pasadena, JPL develops and manages planetary programs and operates the Deep Space Network. The Ranger and Surveyor lunar programs; the Mariner, Viking orbiter, and Voyager planetary programs; and the current Infrared Astronomical Satellite number among the Laboratory's projects.

Johnson Space Center (JSC)

The Lyndon B. Johnson Space Center near Houston is responsible for design, development, and testing of manned flight vehicles; for selection and training of spaceflight crews; ground control of manned flights; and many of the experiments carried aboard the flights.

The lead Center in management of the Space Shuttle program, one of its best known facilities is the Mission Control Center from which manned flights, starting with Gemini IV, have been controlled.

Kennedy Space Center (KSC)

The John F. Kennedy Space Center in Florida is NASA's primary center for the test, checkout, and launch of space vehicles and will be the primary launch and landing site for the Space Shuttle.

Located on Merritt Island near Cape Canaveral, KSC was created to launch the Apollo lunar missions and was used for both Skylab and the Apollo-Soyuz Test Project. KSC also launches a variety of unmanned vehicles from facilities at the Eastern Space and Missile Center, Cape Canaveral, and the Western Space and Missile Center in California.

Langley Research Center (LaRC)

NACA's first laboratory in Hampton, Virginia, has continued as a research center where the development of advanced concepts and technology for future aircraft emphasizes environmental effects, performance, range, safety, and economy. The Center was responsible for the Lunar Orbiter and Viking Mars lander projects and was the home of Project Mercury. It is developing the Long Duration Exposure Facility for use with the Space Shuttle as well as large advanced space systems concepts.

Lewis Research Center (LeRC)

The Lewis Center in Cleveland has continued its NACA activities as a

Aerial view of Ames Research Center

propulsion laboratory—advancing technologies for aircraft propulsion, propulsion and power generation for spaceflight, and space communications systems—and manages two major launch vehicle programs, Atlas Centaur and Titan Centaur. It also manages many of NASA's support of other Federal energy programs. Its specialized facilities include a zero-gravity drop tower and chambers for testing jet engine efficiency and noise.

Marshall Space Flight Center (MSFC)

Known for its role in developing the Saturn launch vehicles, the George C. Marshall Space Flight Center in Huntsville, Alabama, is now responsible for the External Tank, Solid Rocket Boosters, and engines for the Space Shuttle Orbiter. Its staff also developed the Lunar Roving Vehicle, had program responsibility for the High Energy Astronomical Observatories, and now is responsible for Spacelab and the Space Telescope. MSFC also operates the Michoud Assembly Facility in New Orleans and the Slidell Computer Complex, Slidell, Louisiana.

National Space Technology Laboratories (NSTL)

NSTL in Bay St. Louis, Mississippi, is responsible for the static test firing of large space and launch vehicle engines. It also houses a selection of environmental research and Earth resources activities of NASA and other government agencies.

Wallops Flight Facility (WFF)

NACA's Pilotless Aircraft Research Station on Wallops Island, Virginia, became NASA's only rocket flight-test range. It prepares, assembles, launches, and tracks space vehicles

1963 aerial view of Goddard Space Flight Center, the first facility built by NASA.

Mission Control at Johnson Space Center during STS-2, November 1981.

Kennedy Space Center, launch and recovery site for the Space Shuttle: The 525-ft Vehicle Assembly Building dominates the Launch Control Center at right and the Orbiter Processing Facility at left. The landing runway is visible in the background.

from small sounding rockets to the Scout four-stage solid fuel rocket. Its facilities are used for aeronautical research projects from helicopter and aircraft drop tests and noise projects to laser and radar tracking of aircraft.

NASA's Beginning

The civilian space program was formed from a group of space projects being conducted by the Department of Defense—lunar probes, a communications satellite, rocket engine research. A week later the first U.S. man-in-space effort, Project Mercury, was approved.

There followed, first tentatively, then with mounting confidence and success, a wide array of projects. NASA's aeronautical programs were well-known and recognizable, but its new space programs brought a new vocabulary and new objects to daily life.

Spacecraft, Sounding Rockets, Satellites, Space Probes

A spacecraft is any vehicle that operates above the altitudes attainable by research balloons and aircraft—approximately 30,480 m (100,000 ft) of altitude.

Sounding rockets break through the atmosphere into space for only a few minutes. Although they do not linger long at high altitudes, they have made major discoveries in space science.

A satellite is a spacecraft that has been given sufficient velocity by its launch vehicle to be placed in orbit. Ultimately, the trace of atmosphere still present at satellite altitudes will slow the satellite down and gravity will pull it back to Earth.

Spacecraft launched deep into space that escape the gravitational pull of Earth completely are called space probes. Depending on the target, they are called lunar, planetary, or deep space probes, and they have three possible objectives: flyby, orbit around the body, or impact on the surface with either a hard or soft landing. Flyby probes often go into orbit about the Sun after planetary encounter.

Spacecraft have innumerable classifications: manned or unmanned, recoverable or unrecoverable, active or passive. A passive satellite transmits no radio signals, but may reflect them back to Earth; active satellites emit radio signals to make tracking easier and to transmit data from their instruments to ground stations. They are classified by orbits. A polar

Astronauts working underwater in the Neutral Buoyancy Simulator at the Marshall Space Flight Center to evaluate methods of equipment that might be used to service large structural beams in space.

Wallops Flight Facility launch site.

satellite orbits over Earth's polar regions. A synchronous satellite orbits Earth in the same length of time it takes the Earth to make one revolution on its axis. If the synchronous satellite is also an equatorial satellite, it will seem to remain in the same position in the sky at all times and is then a stationary, or geostationary, satellite.

NASA divides satellites into major categories: (1) Scientific satellites, which carry instruments to measure magnetic fields, space radiation, solar characteristics, or telescopes for particular uses; and (2) Application satellites, which forecast weather, survey Earth resources, extend communications.

Names, Letters, and Numbers

The names given to projects and programs originate from no single source or method. Some have their foundations in mythology, legend, and folklore. Some have historic connotations. Some are based on straightforward descriptions of their missions, often resulting in acronyms. Some grew out of a formal process within NASA under the NASA Project Designation Committee; others evolved more casually and were officially adopted after their use had become widespread.

Spacecraft that are part of a series are usually designated with a letter (Pioneer A) before launch and with a numeral (Pioneer 1) after a successful launch.

The First 25 Years

NASA's first quarter century is a story of learning, a combination of science, technology, engineering, and human experience. It is the continued progress of NASA aeronautical research and the development of true aerospace vehicles.

It is the progress from the first small satellite, from the time when only a handful of scientists and engineers believed humans would fly in space through many changes. The present family of spacecraft differ greatly in size, shape, complexity, and purpose. They are small, large; spin-oriented and attitude-controlled; manned or automated robots; some are in low orbits, some on their way out of the solar system; some are in space until they expire, some are commanded to return to Earth. Some are hardly noticed, others engender tremendous excitement.

The following chapters introduce NASA's programs.

For the Classroom

1. NASA is an independent agency. Ask your students to investigate how it differs from, is similar to, other government agencies (National Science Foundation, Federal Aviation Administration, National Oceanic and Atmospheric Administration).
2. Have your students read the myths that suggested names for space projects familiar to them and discuss their aptness; suggest alternate names; suggest names for future projects.
3. Locate the NASA facilities on a map. Ask your students to list the reasons why particular sites were chosen (research the history and geography of the area).

II

Aeronautics

The first A in NASA stands for aeronautics. In 1983 your students accept aviation as an integral part of their lives. Contrast for them air travel in 1958 and now—propeller planes and jet transportation, 70-passenger airliners and jumbo jets that carry over 400 people, the dominant sea travel of 25 years ago and today's regular SST flights across the Atlantic.

Flying is so accepted a part of life that the immense strides made in just 25 years are hardly remembered. Nor is it recognized that most of the advances of these 25 years began with research in NASA's laboratories. The aeronautical research of the National Advisory Committee for Aeronautics (NACA) was assigned to NASA in its charter, including the objectives:

- The expansion of human knowledge of phenomena in the atmosphere . . . ;
- The improvement of the usefulness, perfor-

An X-15 rocket airplane streaks across the Mohave Desert sky leaving a plume contrail after being released from the mother aircraft B-52.

Technicians assist NASA pilot Joseph A. Walker following a record-breaking flight in the X-15, April 30, 1962.

mance, speed, safety, and efficiency of aeronautical . . . vehicles;

- The preservation of the role of the United States as a leader in aeronautical . . . science and technology;
- The most effective utilization of the scientific and engineering resources of the United States in order to avoid unnecessary duplication of effort, facilities, and equipment.

With some of the most sophisticated aeronautical laboratories and flight test facilities, NASA's research has continued that of NACA. At the major aeronautical centers—Ames Research Center (ARC), Dryden Flight Research Facility (DFRF), Langley Research Center (LaRC), Lewis Research Center (LeRC)—NASA scientists, engineers, and test pilots work closely with universities, other government agencies, and U.S. industry in a wide range of programs and projects.

Aeronautics is a many-faceted subject that can be studied from several different approaches. First, chronologically, by investigating the attempts, successes, and failures of would-be fliers through history. Or by discipline—aerodynamics, guidance and navigation, materials and structures, propulsion.

A third method is to study the tools of aeronautical research: mathematical and physical analysis, now largely computerized; wind tunnels; simulators; and full-scale flight research.

Finally, there are the programs themselves. In examining individual projects, history, discipline, and tool come together to provide an overall view of little-known but challenging areas of aeronautical research. The following list is an introduction to aeronautics at NASA—the research subjects, their aims, and their results.

X-15

March 25, 1960–October 24, 1968

The X-15—a 15-meter (50-foot-long), black, stub-winged, rocket-powered flight research craft with a conventional nose-wheel and skids mounted at the rear for landing—was a true aerospace vehicle. With wings and aerodynamic controls it traveled like an airplane in the atmosphere, and in flight beyond the atmosphere, like a spacecraft.

It was launched from beneath the wing of a B-52 at an altitude of 13,716 meters (45,000 feet). After its drop, the rocket engine was fired and the craft climbed in a steep trajectory, then nosed over to descend in a glide to a landing.

Through a series of progressive steps, the X-15 set new altitude (17,960 m or more than 67 mi) and speed (6.7 times the speed of sound) records. Its 199-flight program contributed important data about weightlessness, aerodynamic heat, atmospheric entry, the effect of noise on aircraft materials, and piloting techniques.

The X-15 was a joint NASA/Air Force/Navy project. First piloted by A. Scott Crossfield, both Neil Armstrong, commander of Apollo 11, and Joe Engle, commander of the Shuttle's second flight, were among the pilots who flew the X-15 into unexplored areas of flight.

Supersonic Cruise Aircraft Research (SCAR)

NASA researchers worked throughout the 1960s on technologies for supersonic transport. By 1971, Boeing's Supersonic Commercial Air Transport (SCAT) was ready for production, but concerns about noise, economy, and pollution prevented further funding. Convinced that supersonic transport research would eventually pay off, in 1973 the government funded the Supersonic Cruise Aircraft Research (SCAR) program. Nine years of a sustained, focused technology program involving NASA and major U.S. propulsion and airframe companies resulted in significant improvements over earlier supersonic transport concepts. By the early 1980s, the SCAR program had developed technologies permit-

ting a greatly increased range, greater passenger capacity, lighter weight, and cleaner, quieter, more efficient engines.

Terminal-Configured Vehicle (TCV)

With the continually growing use of air transportation, air terminal problems increased: approach and landing in bad weather, safety and efficiency in controlling high-density traffic, and noise of aircraft in take-off and landing over densely populated areas.

Recently renamed Advanced Transport Operating Systems Program (ATOPS), the Terminal-Configured Vehicle (TCV) is a research tool, a standard Boeing 737 twin-jet transport with a second cockpit in the passenger cabin. Equipped with state-of-the-art instrumentation, the second cockpit is the flight center for the research, while safety pilots fly in the conventional cockpit for backup.

In 1979 the TCV was used to demonstrate the Microwave Landing System (MLS) and Area Navigation in efficient descent and airport approach paths and precision flight control. Its success led to the International Civil Aviation Organization's adoption of MLS as the world standard.

Pivoting Wing

Several decades ago Robert T. Jones, NASA scientist at ARC, invented the concept of an aircraft wing that could pivot up to 60 degrees in flight; years of analysis and wind tunnel tests suggested the results would be considerable fuel economy.

A small, piloted research aircraft called Ames-Dryden-1 (AD-1) was built, and in 1979 made its first flight. During takeoff, landing, and low-speed cruise, the AD-1 flies with wings at right angles to the fuselage. At higher speeds, the wing pivots so that the right half sweeps forward and the left half sweeps back. The pivoted wing decreases air drag, allowing the plane increased speed.

The Ames-Dryden-1 (AD-1) in a flight test of its pivoted wing.

The AD-1 flight research program, completed in 1981, tested the pivoting wing in 39 flights at speeds up to 165 mph.

HiMAT

Highly Maneuverable Aircraft Technology (HiMAT) is a NASA/Air Force flight research program to study and test advanced fighter aircraft technologies.

The HiMAT vehicle is a 44-percent scale model with wing tip-mounted winglets and a small forward canard wing for high maneuverability. It consists of a core design to which modular components can be attached easily and replaced, a format that allows low-cost testing of a variety of concepts.

In 1979 the remotely-controlled research aircraft made its first flight. The following year it achieved near-maximum design maneuverability at sustained near-supersonic speeds, and in 1981 its flight testing was expanded to transonic speeds.

The HiMAT flight test program ended in January 1983. The vehicles

This six-foot diameter experimental turbofan was evaluated in noise tests as part of the Quiet Engine Program at Lewis Research Center.

had performed superbly with maneuverability equal to or above the goals of the design.

Lifting Bodies

Aeronautical research does not often extend to the problems of spacecraft. An aerospace vehicle, such as the Space Shuttle orbiter, to fly in the atmosphere safely, must be aerodynamically stable and maneuverable at hypersonic, supersonic, transonic, and subsonic speeds. Known as a lifting body, this type of craft was researched for many years before its application to an aerospace flight.

NASA has had three experimental lifting bodies, which are wingless and achieve the aerodynamic lift and maneuverability necessary for flight from their body shape alone. The first, ARC's M2, featured a flat top and round belly. The second, HL-10, was developed at LaRC and had a rounded top and flat belly. The third is the NASA/AF X-24. The vehicles were carried aloft by a B-52 and released to glide to landings on a dry lake bed. The X-24B had made 33 successful flights when the program was completed in 1975.

Forward Swept Wing (FSW)

The Forward Swept Wing (FSW) offers the potential for high performance design with both civil and military applications. In a joint program with the Defense Advanced Research Projects Agency, NASA is testing the unusual wing which is swept forward at a 30 degree angle to the fuselage.

Wind tunnel tests, composite element tests, and simulations indicate the FSW design should give greater maneuverability at transonic speeds and superior low-speed performance. To avoid structural deflection of the wing, its design calls for laying up the composite material plys in definite patterns. The X-29A is scheduled for demonstrator flights at the Dryden Facility early in 1984.

Quiet Engine Research

The Lewis Research Center has led the investigation for reducing noise and pollution produced by airplanes. Beginning in the late 1960s, the Quiet Engine program focused on developing an engine with noise levels 15 to 20 PNdB (Perceived Noise Decibels) below levels then in use. The results: (1) a high bypass ratio turbofan engine to help produce thrust with low velocity air; and (2) a retrofittable acoustic nacelle, an engine housing lined with sound absorption material.

Quiet, Clean, Short-haul Experimental Engine (QCSEE)

In the late 1970s, the QCSEE program began testing two research engines at LeRC. One engine is mounted beneath the wing, and the other is designed for placement above the wing. Developed for a Short Takeoff and Landing (STOL) aircraft but applicable to the larger commercial airliners, these engines direct their exhaust downward with wing flaps to add lift for short take-off

Model of a Short Takeoff and Landing (STOL) aircraft in Langley Research Center's Full-Scale Wind Tunnel.

and landing. Tests have demonstrated the engine's ability to operate at a noise level 60 to 75 percent below that of engines now in service. Carbon monoxide and unburned hydrocarbon emissions have also been dramatically reduced.

Quiet, Clean, General Aviation Turbofan Engine (QCGATE)

The QCGATE program was directed toward meeting U.S. environmental standards for general aviation engines. An existing turbojet or turbofan engine core was used in the experimental, quiet high-bypass turbofan engine which incorporated the latest quiet engine technologies. In 1980 the QCGATE program was completed with the resulting research engines producing from 50 to 60 percent less noise than the most quiet current business jets.

V/STOL Research

NASA is developing a number of new flight technologies for safe, clean, quiet, and efficient Vertical and Short Takeoff and Landing (V/STOL) aircraft.

Two VTOL programs, Rotor Systems Research Aircraft (RSRA) and Tilt Rotor Research Aircraft (TRRA), are joint NASA/Army projects. In STOL research, NASA is experimenting with propulsive-lift concepts with the Quiet Short-haul Research Aircraft (QSRA).

Rotor Systems Research Aircraft (RSRA)

The Rotor Systems Research Aircraft (RSRA) is designed to test various advanced rotor systems. Able to fly as a conventional helicopter, the RSRA also flies with wings to assist the lift and is able to operate in a wide range of speeds. The two RSRA currently in use are helping to develop technologies for safer, quieter, more reliable helicopter performance.

Tilt Rotor Research Aircraft (TRRA)

The XV-15 Tilt Rotor Research Aircraft (TRRA) employs two large rotors to combine the advantages of a helicopter's vertical lift with an airplane's cruising speed. In the air, the rotors tilt forward to become propellers for cruising. This versatile aircraft can take off and land vertically, hover, and fly forward, sideways, or rearward.

The TRRA is potentially valuable as a commercial commuter liner operating out of close-to-city heliports. In 1981 the TRRA completed the proof-of-concept flight research phase. It flew twice as fast and twice as far as a helicopter on an equal amount of fuel and achieved a top speed of 557 km/h (346 mph).

Lifting Bodies. Left to right: X-24A, M2-F3, HL-10.

Quiet Short-haul Research Aircraft (QSRA)

An experimental vehicle, the Quiet Short-haul Research Aircraft (QSRA) addresses airport congestion and noise problems. The QSRA has demonstrated the effectiveness of propulsive-lift technology, where the engine's exhaust is directed over the wing surfaces, which increases lift and allows quiet takeoffs and landings from short runways.

In 1981 the QSRA completed a flight evaluation series during which

government, military, airline, and industry pilots flew the aircraft.

A technician at Dryden Flight Research Facility readies the pilot's cockpit of the supercritical wing test aircraft for flight.

Aircraft Energy Efficiency (ACEE)

In response to a U.S. Senate request in 1975, NASA established the Aircraft Energy Efficiency (ACEE) program to develop fuel-saving technologies for both existing and future aircraft. Using an inter-disciplinary approach, ACEE includes six major technology programs to explore ways to improve both engine and airframe performance: more efficient wings and propellers; new composite materials for airframes that are lighter and more economical than metal; ways to make today's jet engine more fuel efficient; new engine technologies for energy-saving aircraft of the future.

Energy-Efficient Transport (EET)

An important factor in flight efficiency is the shape of an aircraft and the resulting flow of air over its surfaces in flight. Developing improved wing designs is a major task of the Energy-Efficient Transport (EET) program.

NASA's supercritical wing is shaped to minimize air drag without loss of lift. It also increases volume for fuel storage while improving structural efficiency of the wing, leading to lower weight. A well-designed supercritical wing can reduce fuel consumption 10 to 15 percent. Further fuel efficiency can be achieved with the use of nearly-vertical winglets installed on the wingtips of aircraft, which help to reduce air drag and produce thrust.

Laminar Flow Control (LFC)

The smooth flow of air over the surfaces of an airplane, called laminar flow, occurs at low speeds. At cruising speeds, however, flow becomes turbulent, causing drag and reduced efficiency. The Laminar Flow Control (LFC) program aims to achieve smooth air flow at cruising speeds. Technology combining the promising concept of lightweight suction systems to remove portions of turbulent air through multiple slots or tiny holes on the wing surface with the new supercritical wing designs is being tested for use on commercial aircraft in the 1990s. The LFC program has combined detailed analysis and model testing in its early phases of research and development. Flight testing, the third phase of the program, is scheduled to extend through September 1986.

Advanced Turboprop (ATP)

Renewed interest in fuel economy has the more fuel-efficient turboprop engine being reconsidered and improved for future use. Odd-looking new multi-bladed propellers are being developed for use on a turboshaft engine. The improved turboprop aircraft is expected to compete favorably with jetliners for speed and noise, but be more fuel efficient.

The three-phased Advanced Turboprop (ATP) program is testing small-scale propeller models to establish proof-of-concept. In the second phase, large-scale propellers will be used to validate structural dynamics, and in the third, a full-scale experimental propeller will be tested in flight.

Engine Component Improvement (ECI)

The Engine Component Improvement (ECI) program objectives were to reduce the cycle of wear and deterioration that affects fuel efficiency of jet engines. ECI developed new components for existing engine designs to resist the erosion, leaking, and warping responsible for efficiency loss; a highly effective seal for the turbine engine to prevent the engine's high-pressure gases from leaking out of the main flow path and remain effective under conditions that cause

conventional seals to fail; new materials or ceramic coatings that can reduce erosion and corrosion of turbine blades; and improved aerodynamic design of the compressor and blades that contributes to engine efficiency.

Major successes of the ECI program were realized early in the 1980s and have become available for use in the new Boeing 767 and the McDonnell Douglas DC-9 Series 80 aircraft.

Energy Efficient Engine (E^3)

The Energy Efficient Engine (E^3) program is planning a completely new engine design for use after 1990. Using the standard building-block technique of engine manufacturers, NASA researchers and engineers refine each new component to develop a core design to which the fan, turbine, and exhaust nozzle are added. The E^3 program is scheduled to complete testing of new components early in the 1980s.

One area of study focuses on increasing the engine's cycle pressure ratio and turbine operating temperature, converting a greater proportion of fuel into energy. Another component mixes the engine's cool bypass air with the hot core stream, increasing propulsion without added fuel. These E^3 components will also help to reduce noise and exhaust pollution.

Composite Materials

Unnecessary weight adds to the amount of fuel needed for flight, so the ACEE program has been developing technolgy for new lightweight composite materials for airframe construction.

Conventional aircraft are constructed primarily with alloys of aluminum, magnesium, titanium, and steel; the new composite materials consist of graphite, glass, or Kevlar® fibers arranged in a matrix, generally epoxy. By arrangement of the fiber orientation, the great strength of these materials can be directed along a line or in random directions. Light, yet strong and stiff, the materials offer possible weight reductions of 25 percent or more. Beginning with secondary structures not critical to flight safety, some new materials have been flight-tested. The goal is to monitor the materials in daily use on a commercial airline, where the normal wear on the pieces can be observed; because they replace metal parts on aircraft in service, each new part will be certified by the Federal Aviation Administration (FAA). Eventual testing of a complete wing and fuselage will provide a design base for future energy efficient aircraft.

A test aircraft is suspended at Langley Research Center's Impact Dynamics Facility for a simulated free flight crash test.

Aeronautical Safety

Today's aircraft incorporate many improvements developed over the years to make them safer for flight in both good and bad weather, and to increase safety during takeoff and landing.

Crashdynamics

Recent studies have included an investigation of airplane crashdynamics information with the intent of increasing the survivability of passengers in an accident. For several years, NASA has been deliberately crashing controlled, extensively instrumented aircraft, both single- and twin-engined.

The planes, containing anthropomorphic dummies harnessed in the crew and passenger seats, are crashed onto a runway from a test rig. The data collected helps researchers understand how an aircraft absorbs the energy of impact and transfers the shock to passengers. The tests include the study of improved seats, harnesses, and crushable sub-floor and fuselage structures.

Fireworthiness

In a related effort, NASA researchers at the Ames and Johnson Centers are developing fire resistant materials for use inside cabins. One concept uses fire resistant wrappings over conventional polyurethane foam cushions. Another fire resistant, lightweight polymide seat cushion has been developed at Johnson and is being evaluated in service by three airlines. Similar lightweight fireworthy materials are being applied to ceiling, wall, and floor panels.

Less flammable jet fuels are also under development, most notably the British-developed AMK safety fuel, FM-9. Full-scale tests have demonstrated the new fuel's ability to prevent major fires caused by ignition of jet fuel during and after a crash. Along with the FAA, NASA has been testing the safety fuel and evaluating its compatability with the most common engine in service.

Automated Pilot Advisory System

For general aviation pilots operating out of small uncontrolled airfields, NASA has developed and successfully demonstrated the Automated Pilot Advisory System (APAS) to provide weather, traffic, and airport information. The APAS includes a tracking radar, weather sensors, a computer, and a transmitter.

Computer-generated voices broadcast traffic information every 20 seconds within three miles of the airport, and every two minutes, information on airport identification, active runway, wind speed and direction, barometric pressure, and temperature.

Stall/Spin Research

The stall/spin phenomenon has been a major cause of accidents in general aviation. A stall occurs when the angle of attack of the wing increases to the point where air across the wing separates instead of following the upper surface; this causes a loss of lift. Following a stall, an airplane sometimes will begin to spin downward at a rapid rate. Stall/spin tests have ranged from early studies with models in wind tunnels and special spin tunnels to more recent use of simulators and full-scale flight research vehicles.

In the 1970s a large-scale effort focused on vertical tail designs and went on to develop a number of leading-edge wing extensions. These extensions have been shown to make test airplanes significantly more resistant to spin.

The stall/spin research has produced a large body of data that aids industry in the design of safer airplanes.

Icing Research

An increasing demand for all-weather flights brought on by advances in avionics systems, has brought a renewed interest in improving aircraft performance under icing conditions. Current research is aimed toward developing lightweight, low-power consumption, cost-effective ice protection systems. Analysis, wind tunnel testing, and flight research are being used to validate the effectiveness of these protection systems.

In 1982, NASA developed a long term icing research program in cooperation with the Army, Air Force, FAA, and the governments of Canada and Great Britain to evaluate icing instrumentation that had been tested at Lewis. The Center also initiated research on protection systems for airfoil leading edges, using an electro-impulse concept. NASA also provides the FAA with icing research data to support upgraded aircraft certification, particularly for rotorcraft.

Aviation Safety Reporting System

In cooperation with the FAA, NASA completed in 1982 the development of the Aviation Safety Reporting System (ASRA), a voluntary, confidential, nonputative reporting system designed to surface deficiencies in the National Aviation System before accidents occur. Since April 19, 1976, the System has received more than 30,000 reports, issued 740 alert bulletins, and published 240 reports.

For the Classroom

1. Research topics:
   - The uses of general aviation
   - Compare a large metropolitan airport and a small general aviation airport
   - Airport terminals—the early structures, contemporary complexes, airports of the future
   - How the local airport, or lack of one, affects a community
2. Plan a field trip to your local airport.
3. Have students list as many types of aircraft as they can, their characteristics and their uses. How are they alike? different?
4. Have your students research Reynolds and Mach numbers; differentiate between subsonic, supersonic, transonic, and hypersonic speeds.
5. The difference between laminar and turbulent flow can be easily demonstrated with a burning piece of punk or stage cigarette in an ash tray; note the smooth flow upward which abruptly changes to turbulent. The same effect can be shown with a stream of water from a faucet. The point at which the flow changes from laminar to turbulent is at the Reynolds number. To show how an aircraft flies, i.e., the flow around the wing, one can demonstrate the coanda effect by placing one's finger (or a test tube) in the water flow.

III

Applications Satellites

When the first satellites were launched in the late 1950s, many people were skeptical about the practical value of a space program. In just three years, however, observations and measurements from Earth-orbiting satellites were revolutionizing communications and weather forecasting and showing Earth on a global scale. These were the applications satellites, spacecraft with experiments and instruments that provided unique, direct benefits to life on Earth. They and those that evolved from them have made it possible for people on opposite sides of Earth to communicate instantaneously, for people in remote areas of the world to learn by television, for ships to know where storms and icebergs threaten passage, for forecasters to watch weather develop, for oil companies to locate drilling sites, for environmentalists to monitor the spread of pollutants.

In both domestic and foreign applications satellite

Many well-known landmarks in New York City and its environs are visible in these 30-meter resolution Thematic Mapper images taken from Landsat 4.

programs, NASA has contributed research and development, launching capabilities, and evaluation of spacecraft. The technologies developed produced passive and active communications satellites, the first synchronous and geostationary orbits, and the cloud cover pictures that now are a regular feature of daily weather reports.

From Echo, the balloon that was the first satellite everyone could see, and the scientific Explorer 6 that also took the first crude cloud cover picture, the applications satellites have become complex multipurpose systems.

Once NASA has developed the weather and communications satellites, the responsibility for operating them falls to other government agencies or to private industry. NASA continues its research role, seeking and developing advanced technologies.

The following list introduces the major groups of these satellites, their purposes, and the benefits they have contributed.

Communications

In 1945 British scientist and science fiction writer, Arthur C. Clarke, published a technical paper in which he suggested that communications satellites were feasible. Fifteen years later, NASA launched its first communications satellite, Echo, a silvery balloon that orbited Earth every 114 minutes.

Echo was a passive satellite that reflected radio signals back to Earth. Two years later, Relay, the first active satellite was launched to receive signals, amplify them, and transmit them back to Earth.

Today's split-second global communications by voice, television, and computer are such a part of daily life that the evolution from simple passive reflectors to complex active transmitters is hardly remembered. After NASA completed research and development, private companies produced their own communications satellites, and in 1962 Congress authorized the Communications Satellite Corporation, Comsat, which is the U.S. representative in and manager of Intelsat, the International Telecommunications Satellite Organization. For both industry and Intelsat, NASA launches and tracks satellites on a cost-reimbursable basis.

Echo

The Echos were inflated in space to spherical balloons of aluminized Mylar, 30.5 and 40 meters (100 and 135 feet) in diameter, respectively. Passive communications satellites, they reflected radio signals between ground stations. They also provided information about the density of the upper atmosphere. Echo 1 was launched August 12, 1960, Echo 2 in January 1964.

Relay

Relay 1 was NASA's first active repeater experimental satellite; launched December 13, 1962, it handled 12 simultaneous two-way telephone conversations or one television channel and provided the first satellite communications link between North and South America and Europe. Relay 2, an improved version, was launched in January 1964.

Syncom

Three experimental, active satellites; the name, coined from the first syllables of "synchronous communications," referred to their orbits. Weight: 38.5 kg (about 85 lbs) each.

Syncom I
February 14, 1963
In nearly synchronous orbit, but communications failed.

Echo.

Relay 1, NASA's first active repeater satellite, was an eight-sided prism 33″ high & 29″ in diameter at its broad end. The exterior honeycomb aluminum panels were studded with 8,215 solar cells.

Syncom II
July 26, 1963
First satellite placed in synchronous orbit. Many successful intercontinental communication experiments.

Syncom III
August 19, 1964
First stationary Earth satellite. Demonstrated the practicality and effectiveness of stationary,

The Applications Technology Satellite (ATS-6) is shown in final systems test and checkout at the Fairchild Industries Plant, Germantown, Maryland.

active communication satellites. In orbit near the International Date Line, it was used to telecast the 1964 Olympic Games in Tokyo to the United States, the first television program to cross the Pacific.

Applications Technology Satellites (ATS)

A series of six multipurpose Applications Technology Satellites designed to test new space instruments and demonstrate new satellite technologies, particularly those used in synchronous orbit satellites.

ATS-1
December 6, 1966
Took first U.S. high-quality photographs of Earth from synchronous orbit, showing changing cloud-cover patterns. Also relayed color television across the U.S. and was the first satellite to permit two-way VHF communication between ground and aircraft in flight.

ATS-3
November 1967
Carried advanced communications, meteorology, and navigation experiments; transmitted color images of one complete side of Earth.

ATS-6
May 1974
The first communications satellite with power to broadcast TV photos to small local receivers; also used for a number of experimental public health and education telecasts to remote rural areas in the U.S. and India. (See Chapter IV, India.)

Communications Technology Satellite (CTS)

January 17, 1976
The CTS was a joint project with Canada. A high-powered satellite, it used a movable terminal to investigate the possibility of transmitting public service information to small, inexpensive antennas in remote locations.

Earth Resources

Earth observation satellites have brought us a new view of our planet. Mountains, prairies, deserts, lakes, rivers, reservoirs, forests, farms, cities, highways, have become infrared and ultraviolet scenes. Millions of these pictures have been distributed to users of Earth resources information around the world.

From the outset, the remote sensory devices of these spacecraft have produced a continuous flow of data. The results, including often dramatic pictures, have been tangible and the satellites unique tools of enormous practical value for a wide range of interests: urban development and land use and water resource management, agriculture, locating pollution, geology, forestry, mapping and charting.

Geologists use the data to locate drilling sites, to predict earthquakes, and to study volcanoes.

Skilled photointerpreters among agriculturists can readily distinguish among a variety of crops in the satellite images. With computers, maps can be produced showing the precise location of each crop over large areas of land. Using this technology, NASA participated in a three-year experiment to monitor global wheat production beginning in 1974. The Large Area Crop Inventory Experiment (LACIE) successfully tested several techniques for predicting crop production early in the growing season.

First coast-to-coast color photomosaic of the United States made from 569 virtually cloud-free images taken by the Landsat 1 satellite orbiting at a height of 570 miles (above).

Landsat 4 (right).

The cartwheel TIROS meteorological satellite, (far right) which provided near global coverage daily, viewed Earth from the sides of the spacecraft rather than from the bottom.

Much of the everyday disposal of tons of garbage and trash and toxic wastes dumped into the environment ends up in our rivers, lakes, and oceans. The challenge to clean up polluted areas and to protect those areas yet untouched requires information on a scale that was unavailable before satellites. Earth resources spacecraft have provided valuable surveys of large areas of land, helping scientists and environmentalists trace the sources of pollution and monitor the dissemination of waste.

Wise management of the Earth's water resources is necessary for both present and future generations. Data from satellites has been helping hydrologists to predict floods and estimate flood damage, as well as to monitor water supplies.

From the simple PAGEOS balloon of 1966 to the advanced Landsat 4 of 1982, the Earth resources experiments have changed radically the way we see Earth, collect information about it, and interpret the results.

Passive Geodetic Earth Orbiting Satellite (PAGEOS)

June 1966

A large metalized balloon, 30 meters (98.4 ft) in diameter, similar to the Echo satellites. A passive satellite, it reflected sunlight and, photographed by ground stations around the world, established a worldwide triangulation network to map Earth's surface.

Landsat

A series of satellites that have provided a wealth of observations which have improved our ability to monitor and understand the dynamics and character of the various features and materials covering the surface of the Earth.

Landsat 1, July 1972

Landsat 2, January 1975

Landsat 3, March 1978

The first Landsats (1 and 2 were originally called ERTS for Earth Resources Technology Satellite) carried an Earth-viewing sensor called Multispectral Scanner (MSS), a radiometer that obtains imagery of Earth's surface in four discrete spectral bands. The decade of their image-collecting showed the unique types of data that MSS imagery could provide—vegetation types, bare soil and rock conditions, snow cover—on a highly repetitive basis. The images Landsats 1, 2, and 3 collected represent the first historical record of Earth's global surface conditions.

Landsat 1 was removed from service in 1978, Landsat 2 in 1982, and Landsat 3 will be retired in 1983.

Landsat 4, July 1982

In addition to the MSS, Landsat 4 has a more sophisticated sensor, the Thematic Mapper (TM) which measures the intensity of surface radiation in seven discrete bands and has approximately twice the spectral resolution, three times the spatial resolution, and four times the sensitivity of the MSS. From a 695-kilometer (432-mi) orbit, it is providing extraordinary details, and for the first time, natural color images, of Earth's surface features.

NASA has transferred the operation and management of Landsat to the National Oceanic and Atmospheric Administration (NOAA). Management control over the TM will be retained during the experimental research and development phase of the new sensor system; NASA expects to transfer control of the TM to NOAA in early 1985.

Landsat imagery is available for a lab service charge. For information about ordering pictures, write to the EROS Data Center, Sioux Falls, SD 57198.

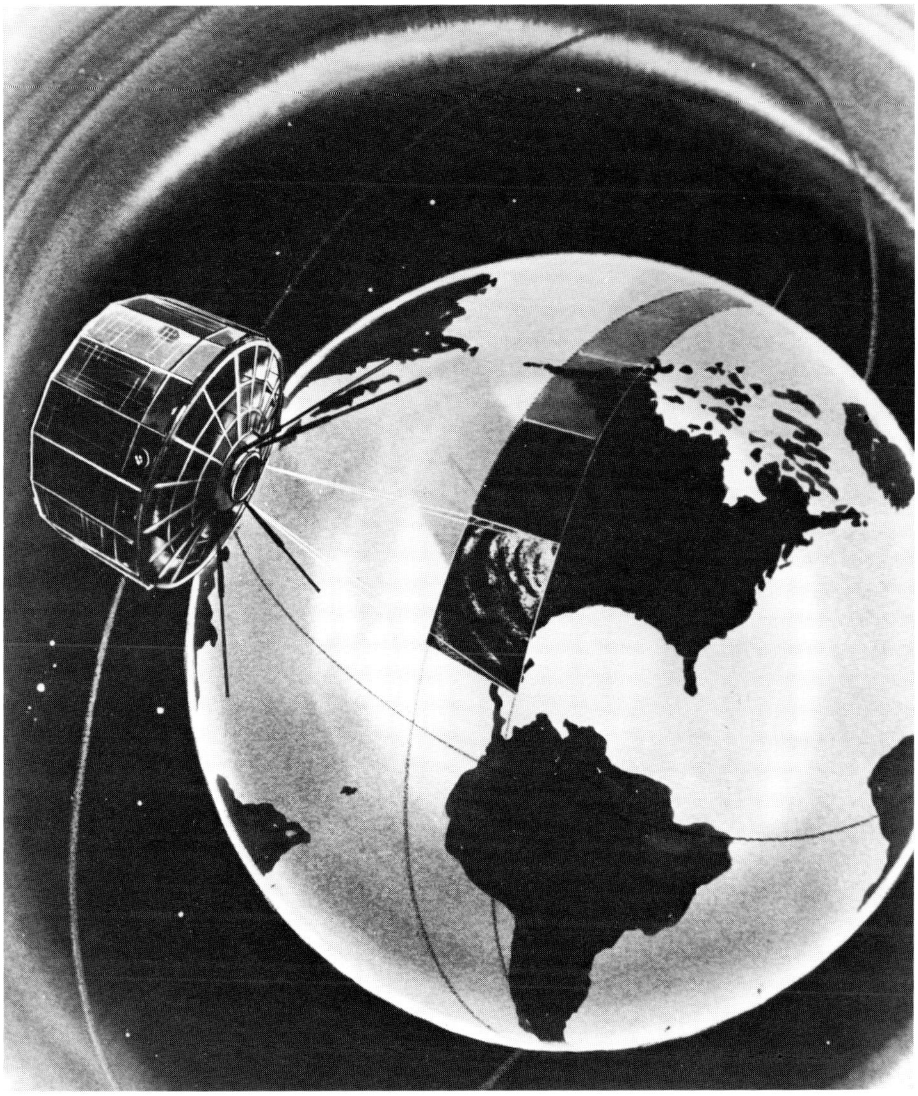

Earth Resources Experiment Package (EREP)

SKYLAB, May 1973–February 1974

Objectives: To test the use of sensors operating in the visible and infrared portions of the spectrum, to test a complex microwave sensor that provided a space-based radar system for Earth resource studies, and to develop data analysis techniques.

Investigations: Agriculture, range, and forestry; land use and cartography; geology and hydrology; oceans and atmosphere.

Results: Demonstrated the potential and practicality of using quality photos from orbiting spacecraft for large geographic as well as regional and local areas and their usefulness as a tool for professionals concerned with management of resources.

Laser Geodynamics Satellite (LAGEOS)

May 1976

A heavy sphere, 411 kg (906 lbs), 60 centimeters (2 ft) across and covered with laser reflectors, designed to demonstrate the feasibility and utility of a ground-to-satellite laser system to contribute to the study of solid-Earth dynamics; provided valuable data to scientists analyzing conditions leading to earthquakes.

Heat Capacity Mapping Mission (HCMM)

April 1978

First in a series of small experimental satellites designed for the Applications Explorer Missions. Later called AEM-1, it had one sensor for one purpose, making thermal measurements of Earth's surface and atmosphere across the U.S. Its unique sensor could read daytime temperatures associated with the Sun and nighttime temperatures associated with radiative cooling.

Meteorology

Weather affects everyone—food supplies, travel, recreation—and along with other applications satellites, the weather satellites have brought special advantages to life on Earth. They enable people to plan ahead, assist meteorologists with forecasting, and help scientists to understand better the air around us.

Advance knowledge of weather systems that can be disastrous is the most striking advantage; part of that knowledge comes from the ability to see the sparsely populated regions of the world where weather is born, thus aiding long-term prediction. For local meteorologists, daily photographs show how their local weather patterns fit into the overall picture.

On April 1, 1960, TIROS 1, the first true weather satellite, was launched. With each succeeding generation of satellites, remote sensing instruments became increasingly sophisticated and today's high quality pictures are a far cry from the first tentative trials.

TIROS

The Television and Infrared Observation Satellite (TIROS) was a simple hatbox-shaped craft carrying special television cameras that viewed

(Far left) The first infrared (temperature) data acquired by the Applications Explorer Mission 1—the Heat Capacity Mapping Mission covering an area approximately 700 kilometers (434 miles) wide, running from south of Cape Hatteras, N.C. to Lake Ontario. Temperature values have been color-coded so that cold to hot is represented by the sequence of purple, blue, green, brown, yellow, orange, red, gray, and white. Black areas at the upper left represent cold clouds.

Skylab 4 EREP photo of San Francisco, Sacramento River, Oakland, and Concord, California photographed from Earth orbit (left).

LAGEOS (left below), looking like a cosmic golfball, provides a stable point in the sky to reflect pulses of laser light.

Earth's cloud cover from a 725-km (450-mi) orbit. The pictures radioed back to Earth provided meteorologists with a new tool—a nephanalysis, or cloud chart.

By 1965, nine more TIROS satellites were launched. They had progressively longer operational times, carried infrared radiometers to study Earth's heat distribution, and several were placed in polar orbits to increase picture coverage over the first TIROS in its near-equatorial orbit.

TIROS 8 had the first Automatic Picture Transmission (APT) equipment that allowed pictures to be sent back right after they were taken instead of having to be stored for later transmission. Eventually, APT pictures could be received on fairly simple ground stations anywhere in the world, even in high school classrooms.

TIROS 9 and 10 were test satellites of improved configurations for the Tiros Operational Satellite (TOS) system. (When it became part of another acronym, TIROS was written Tiros.)

Operational use started in 1966. In orbit, the TOS satellites were called ESSA for the Environmental Sciences Services Administration, the government agency that financed

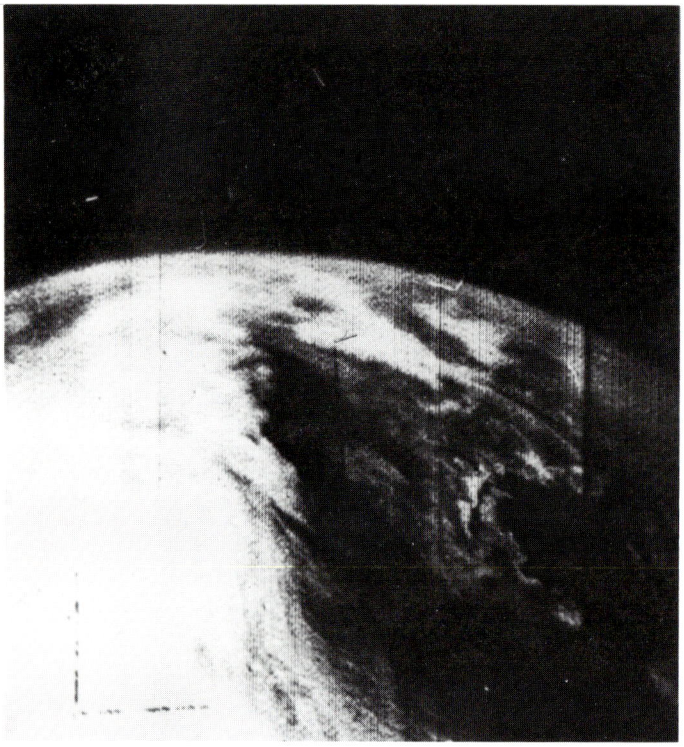

The first weather picture from space (top) by TIROS 1, April 1960, and a view by TIROS-N, April 1980, a third generation advanced meteorological satellite.

A meteorologist (above) examines a sequence of cloud images from ATS-3.

Artist's concept of the Nimbus weather satellite in Earth orbit (right).

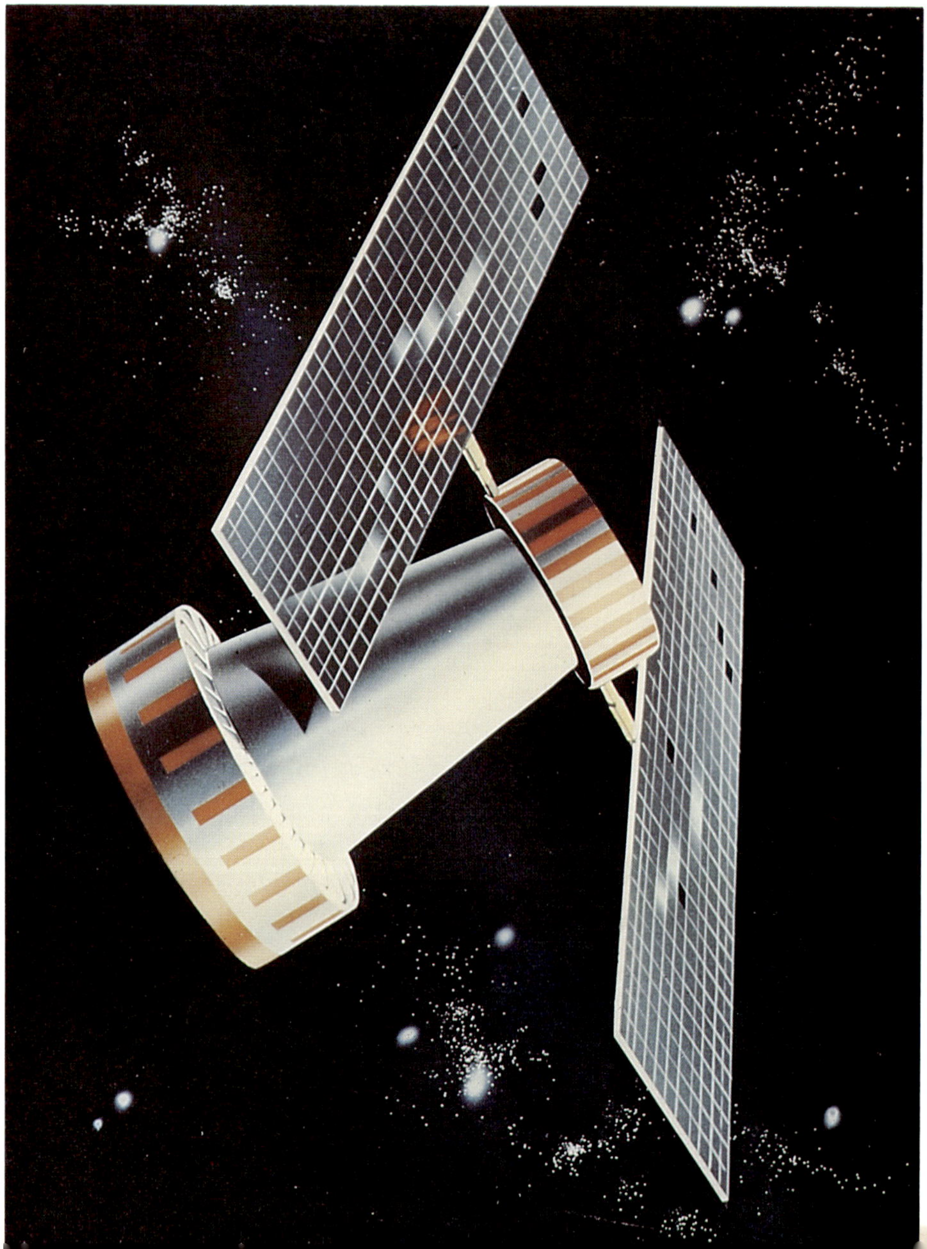

Second in a series of weather satellites, the Synchronous Meteorological Satellite 2 is prepared for launch.

and operated them. TOS satellites were placed in Sun-synchronous orbits, so they passed over the same position on Earth's surface at exactly the same time each day; this allowed meteorologists to view local cloud cover changes on a 24-hour basis.

Several ITOS (for Improved TOS satellites) have been launched since 1970 and are the workhorses of the meteorologists. In orbit they are called NOAA for the National Oceanographic and Atmospheric Administration which is responsible for their operation.

Nimbus

More complex than TIROS, Nimbus was a second-generation research satellite. Each carried advanced cameras, an APT system, an advanced TV cloud mapping camera system, and an infrared radiometer that allowed pictures at night for the first time.

Seven were placed in orbit between 1964 and 1978. Nimbus 3, launched in April 1969, provided data for the U.S. portion of the Global Atmospheric Research Program (GARP), an international program formulating and coordinating research for achieving long-range global weather forecasting.

The Nimbus satellites tested space-borne meteorological equipment and their experiments led to operational, 24-hour satellite weather coverage.

Applications Technology Satellites (ATS)

Intended primarily for communications technology, these multipurpose spacecraft contributed much to advance weather forecasting.

ATS-1
December 1966
Took repetitive photographs of the same area, greatly aiding in the early detection of severe storms.

ATS-3
November 1967
Recorded the first color images of the full Earth disc. Took photos every 20 minutes enabling meteorologists to put them together in a sequence and make a motion picture of cloud movements; until 1975, the cloud cover pictures seen on TV came from this satellite.

Synchronous Meteorological Satellites (SMS-1 and 2)

May 1974 and February 1975
First experimental craft for a geosynchronous satellite system designed specifically to provide weather data and to serve as prototypes for later operational satellites funded by NOAA. Following launch and check-

out by NASA, SMS-1 and SMS-2 were transferred to NOAA for use in the National Operational Meteorological Satellite System.

Successive satellites, designated GOES (Geostationary Operational Environmental Satellite), were constructed and launched by NASA, funded and operated by NOAA.

Oceanography

Seventy percent of Earth is covered by oceans. These vast areas of water are a source of energy in the form of weather, the home of great schools of fish, a mechanism for the disposal of waste products, and the major means of transporting the goods of the world by ship.

Precise knowledge of the oceans' resources and dynamics has potential application in many scientific and commercial pursuits—ship design and port development, fishing, weather forecasting, environmental science, shipping, selection of sites for off-shore drilling. Satellite observations have contributed to our understanding with accurate measurements of surface wind speeds and directions, temperatures, wave heights, and tides and currents; the data have helped to detect storms, map the ocean floor, and monitor the movement of icebergs.

Earth Resources Experiment Package (EREP)

Skylab, May 1973–February 1974
A collection of instruments with relatively low-resolution, middle-spectrum imaging sensors, EREP proved the feasibility of remote-sensing of wind conditions, surface temperatures and roughness, and the recording of visible phenomena, and advanced the study of the interaction of the atmosphere and land and ocean surfaces. Improved versions of the instruments were built for GEOS-3 and Seasat.

Geodynamic Experimental Ocean Satellite (GEOS-3)

April 1975
Measured the changing shape of the oceans' surface, tides, and currents to improve the geodetic model of Earth and knowledge of Earth-sea interactions. Third in the series of Geodetic Earth Orbiting Satellites (GEOS), GEOS-3 was renamed Geodynamic Experimental Ocean Satellite to emphasize its specific mission in NASA's ocean physics program while retaining the GEOS acronym.

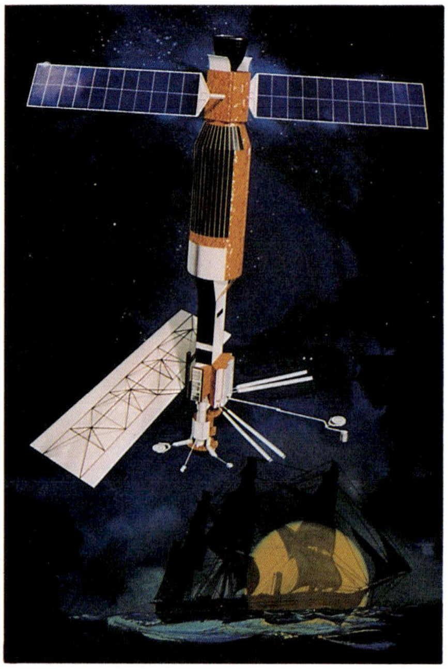

Seasat (Specialized Experimental Applications Satellite)

June 26, 1978
First satellite for sole study of the oceans in a proof-of-concept mission. Objectives: To demonstrate techniques for monitoring Earth's oceanographic phenomena and features from space on a global scale; to provide oceanographic data in a timely fashion to scientists and commercial users; and to determine the key features of an operational ocean monitoring system.

With all-weather and day-night capability, it circled Earth 14 times a day and crossed 95 percent of the oceans' surface every 36 hours giving oceanographers their first worldwide observation of the seas.

Although contact was lost in October 1978 and the mission terminated in November, the objectives were largely met.

Geodynamic Experimental Ocean Satellite (GEOS-3) (above).

Seasat superimposed over an image of the ocean vessel, HMS Challenger, whose round-the-world voyage over 100 years ago became the model for oceanographic voyages (left).

For the Classroom

1. Research topics:
   - History of communications
   - Commercial satellites
   - The development of Earth resources satellites
   - Sources of pollution in the atmosphere
2. What advantage does geological study from space have over study from Earth's surface? from Earth over study from space?
3. Why is a study of the atmosphere important?
4. Secondary school teachers may obtain a copy of *Teachers' Guide for Building and Operating Weather Satellite Ground Stations* from the Educational Programs Officer, NASA Goddard Space Flight Center (202.3), Greenbelt, MD 20771. The publication gives the information needed to construct, modify, and operate a weather satellite recording station.
5. Have your students list the possible benefits of Earth resources satellites; which are apparent in their local community? their state?

DOE
NASA

IV

Energy Research

Meeting the challenges of aerospace exploration brought extraordinary advances in science and engineering and enabled technology personnel to develop expertise in many disciplines. Several years ago it was recognized that NASA's capabilities, developed for aeronautical and space programs, were potentially useful for some areas of energy research and development (R&D). In 1975 NASA and the Energy Research and Development Administration (ERDA)—now the Department of Energy (DOE)—established a working relationship and identified the types of support NASA might provide.

Thus NASA assumed an active role in the national energy R&D program with activities ranging from wind turbines to systems studies and with assistance in developing new power sources and more efficient use of fossil fuels. NASA's major energy programs, which are being conducted at the

Wind turbine on Block Island, Rhode Island. The turbine's blades, 37.5 meters (125 ft) tip to tip, convert wind energy into 200 kilowatts of electric power to supply up to 15 percent of the island's electricity.

Lewis Research Center (LeRC), Jet Propulsion Laboratory (JPL), and Marshall Space Flight Center (MSFC), are in the following areas: Wind Energy, Photovoltaics, Solar Heating and Cooling, Advanced Ground Propulsion, Stationary Power, and Energy Conversion Systems.

Wind Energy

Windmills, sailing ships—picturesque examples of a clean, replenishable source of energy, wind energy, used since ancient times. And for the past 50 years in many countries, it has been a means of generating electricity.

Renewed interest in wind power has been focused in DOE's Wind Energy Program which aims to develop large scale, reliable, cost-effective wind turbines, with operational lifetimes of 20 to 30 years.

NASA has a major role and built the first turbine in 1975 to collect research data; designated MOD-O, it is an experimental 100-kilowatt (kw) machine with a rotor diameter of 38 meters (125 ft) that continues to provide research and engineering data for the design of larger machines. An uprated 200-kw version, the MOD-OA, operated in four locations between 1977 and 1982, providing experience in operating wind turbines with electric utility networks. A larger 61-m (200-ft diameter), experimental machine, called MOD-1, was installed and tested at Boone, North Carolina. Three 2.5 megawatt MOD-2 machines have been operational near Goldendale, Washington, and an advanced MOD-5 design is under development. The newer designs employ larger rotors, which operate at lower wind speeds and could be used in many locations throughout the United States. A single MOD-2 turbine could power 1,000 average sized homes.

The Papago Indian village of Schuchuli in Arizona became the first solar electric community in 1978. The 3,500 watt solar cell system provides energy for lighting homes, powering refrigerators, and running a communal water pump and washing machine.

In another project, NASA has been working with the Department of the Interior (DOI) developing megawatt-size wind turbines to provide power in conjunction with hydroelectric plants. Two machines, a WTS-4 and a MOD-2, were installed near Medicine Bow, Wyoming and dedicated in September 1982. The WTS-4, producing 4 megawatts of electric power, is the most powerful in the world today. Data from these experimental turbines will be collected for two years. If the concept is successful, DOI will consider construction of a wind farm of up to 40 machines.

Photovoltaic Development Projects

The Sun is our most constant source of energy, but utilizing that energy is difficult and costly. Pho-

tovoltaic or solar cells are used to convert the Sun's light into electricity. Solar cells have powered most of NASA's spacecraft, but high costs have made them impractical for industrial, residential, or commercial use. A program at JPL focuses on developing low-cost, long-life solar cells for widespread use on Earth. Substantial progress has been made in improving quality, efficiency, and cost-effectiveness of solar cell module designs.

Several photovoltaic demonstration projects involving NASA and DOE were undertaken during the past decade. They ranged in size from small photovoltaic-powered insect traps to a 3.5 kilowatt system that made the Papago Indians in Schuchuli, Arizona the world's first solar community.

NASA has also worked with the Agency for International Development (AID) on several projects. The first of these was a 3.6 kilowatt solar power system at Tangaye, Upper Volta. The latest was a village power system and a solar-powered drip irrigation system installed in a small village in Tunisia. The photovoltaic power system provides 30 kilowatts of electricity at peak. A variety of other projects, including development of five photovoltaic powered medical systems, is underway for installation in Guyana, Ecuador, Kenya, and Zimbabwe.

Thermal Energy

Direct Solar Heating and Cooling

The Sun's power can be captured, stored, and used to heat and cool homes, offices, and industrial buildings, but technology for solar heating and cooling has been, like photovoltaic systems, expensive and impractical. In an effort to stimulate growth in the solar heating and cooling industry and aid development of affordable, efficient systems,

The reflective surfaces of these parabolic dish concentrators collect and focus sunlight for generating electrical power or high temperature industrial-use process heat. A JPL-DOE project, testing was conducted at the NASA Parabolic Dish Test Site, Edwards, California.

NASA joined the DOE and industry in a number of demonstrations between 1974 and 1981. One project was the Kaw Valley State Bank in Topeka, Kansas, which is completely heated and cooled by solar panels on the roof. NASA monitored the performance of the equipment and analyzed the results. Solar heating and cooling systems have been installed on 48 Federal buildings and all have performed well. NASA's role in these projects ended in 1981.

Solar Thermal Electric Conversion

In another project for DOE, JPL uses large parabolic reflecting dishes to focus sunlight for generating electricity. The reflectors, which are made of silvered glass or aluminized fiberglass, concentrate sunlight onto a heat source for an efficient heat engine. The engine powers an electric generator. The parabolic dish and engine will be combined for use in the Small Community Experiment in Osage City, Kansas.

Advanced Ground Propulsion

Ground transportation vehicles—automobiles, trucks, buses—consume enormous amounts of gasoline and oil every year. As world supplies of petroleum decrease, new automotive technologies will be needed. In a number of joint NASA/DOE projects, research using alternative fuel sources and engine systems has been underway for several years.

A major effort is being directed toward new designs of automotive heat engines, which not only would be more efficient but could use fuels like gasoline, coal, or kerosene. The majority of the work is directed toward two concepts: gas turbine and Stirling-cycle engines. Both engines are potentially fuel efficient, clean, and capable of burning a variety of fuels, including methanol and ethanol.

The research on automotive gas turbines NASA is conducting requires the use of ceramic components to achieve required engine efficiency and low cost. Two parallel development efforts have been underway since 1979, and ceramic component advancements are now being tested in rigs prior to their evaluation in test engines.

The Stirling cycle engine, designed by United Stirling of Sweden, employs an advanced automotive propulsion system capable of a significantly lower fuel consumption than the internal combustion engine. By 1982, four engines had been built and extensively tested. The program now focuses on developing components and materials that will provide a cost competitive engine.

Recently a smaller effort has been started to develop the technology for a high temperature diesel engine for trucks. Much of the materials technology and improvements in aerodynamics of small turbomachinery being developed in the turbine program is expected to be useful in the high temperature diesel. Operation at the higher temperature will improve the engine efficiency and broaden the fuel tolerance of the diesel engine.

In another effort to develop automotive alternatives, a joint NASA/DOE program was authorized in 1976 to develop electric-powered vehicles for widespread use. Current electric vehicles have limited range and poor acceleration and speed performance. NASA's role since 1977 has been to develop technologies for propulsion. Working with industry, NASA is scheduled to complete its development efforts on advanced propulsion components and systems in 1983.

This experimental bus is powered by an advanced automotive gas turbine which provides advantages in fuel efficiency, reduced emissions, lower noise levels, and less dependence on petroleum products. The bus was one of four used on the Washington-Boston route in a two-year experiment.

Stationary (On-Site) Power

While electric utilities currently supply most of industry's electrical power needs, auxiliary boilers fired by premium fuels are utilized on site to provide processing heat. Significant energy savings can be achieved with an approach called cogeneration, in which one power generation on-site system would simultaneously provide all of an industrial plant's electrical and thermal power needs. Cogeneration systems offer the potential for using up to 80 percent of the available energy in the fuel. A number of technologies for industrial cogeneration systems and advanced electric generating systems are being developed.

Gas Turbines

NASA has been advancing the technology for gas turbines for use in industrial cogeneration systems. Turbines designed to burn heavy oils and coal-derived liquid and gaseous fuels efficiently at high temperatures and with acceptable emissions are being tested. NASA will complete its research efforts in 1983.

Fuel Cells

Fuel cells have been important power sources for spacecraft and also offer potential for significant energy savings in cogeneration systems. Fuel cells convert a hydrogen-rich fuel and oxygen (from air) into electricity, providing hot water and steam as by-products in the process. They are clean, quiet, and efficient, and could be used to power large residential and commercial complexes while providing usable thermal energy for heating and air conditioning.

In support of DOE's Phosphoric Acid Fuel Cell Systems Program, NASA is advancing the technology base to allow early development of efficient, cost-effective fuel cells with extended lifetimes, for both multi-megawatt electric power plants and smaller multi-kilowatt cogeneration systems for dispersed residential and industrial use.

Energy Conversion Systems

Magnetohydrodynamics

Magnetohydrodynamics is an advanced coal burning energy conversion process for generating electricity by passing a high temperature gas through a strong magnetic field. DOE studies of this potentially efficient method were supported by NASA from 1977 to 1981 in areas of systems engineering and modelling, critical components assessment, and experimentation.

Coal Gasification

Another potential on-site power generation system uses coal gasification. This concept offers a way to turn polluting high-sulfur coal into a clean-burning gas and to provide usable thermal energy for an industrial cogeneration system. The coal gasification process begins in a gasifier—something like a giant pressure cooker—where treated coal reacts with heated oxygen to produce a fuel gas. Until 1981, NASA provided system engineering support for commercial development of the nation's first coal gasification engineering development facility.

Advanced Coal Extraction

New automated processes for cutting and transportation of coal in an underground mine have been subjects of research since the mid-1970s. Extraction systems must be suitable for resources available after the year 2000, and must promise substantial improvements in production cost and miner safety. NASA has also developed advanced sensors for partially automating current longwall mining machines. These sensors use radar and gamma rays to detect the depth of coal available. One such device, the Natural Background Sensor, has been tested extensively underground. A final report on a system of automated mining components was submitted to DOE in 1982.

Space Utilization Systems

Between 1977 and 1981, NASA and DOE explored the possible development of satellite power stations to provide large amounts of energy to cities on Earth. They concluded that development of such a system was unjustified at the time, but that the possibility for future studies should be left open.

For the Classroom

1. Research topics:
   - Wind power through history
   - Wind turbines
   - Scientists and solar cell research
   - Solar energy
   - Automobile engines of the future
2. Have your students investigate local uses of solar energy. How does it (or might it) affect architecture?
3. Purchase solar cell kits from a local hobby shop, electrical supply store, or other retail store. Let the class build and operate cell-powered models or lights.
4. Have students research solar cell uses and demonstrate them (a possible science fair project).

spacelab

V

International Programs

International cooperation is an important dimension of NASA's program. The agency's 1958 mandate recognized this importance and stated that it "may engage in a program of international cooperation in work done pursuant to this Act, and in the peaceful application of the results thereof. . . ."

During NASA's first year, the United States invited foreign scientists to propose experiments for launching by the new agency. In the 25 years since, an extensive program of international cooperation involving more than 1000 projects with over 100 countries has opened the entire range of space activities to foreign participation, has demonstrated the many peaceful purposes and applications of space science and technology, and has provided opportunities for contribution by scientists and agencies of other countries.

NASA's international cooperation contributes to the U.S. aeronautical and space research program and

The STS-9 orbiter will carry aloft Spacelab 1, shown under construction.

to broader national objectives by:

- developing cost-sharing and complementary space programs;
- stimulating scientific and technical contributions from abroad;
- enlarging the potential for the development of the state of the art;
- providing access to foreign areas of geographic significance for tracking and contingency landing sites;
- enhancing satellite experiments with foreign scientific supporting data;
- extending ties among scientific and national communities; and
- supporting U.S. foreign relations and foreign policy.

The programs fall into two categories: cooperative and reimbursable. The cooperative activities, ranging from flight of foreign-built spacecraft to ground-based study and analysis of data, include contributions of experiments on payloads to be flown in space by NASA, joint projects to develop flight hardware, analysis of lunar samples or data provided by NASA satellites, training, visits, and joint publication of scientific results.

NASA also provides on a reimbursable basis services for which the user country pays; these range from space launch services to data and tracking services.

NASA maintains a number of foreign tracking stations overseas for both the Space Tracking and Data Network (STDN) and the Deep Space Network (DSN), which have been indispensable for the acquisition of data from NASA's many scientific and applications satellites. The host countries have provided sites and personnel for these stations, a cooperation greatly appreciated by NASA.

Ten countries have established Landsat (See Chapter III) receiving, processing, and data distribution facilities and many benefited from the application of Landsat data during the ten years that NASA managed the program.

Automatic Picture Transmission (APT) capabilities which allow local readout of meteorological satellite data, are built into all operational meteorological satellites developed by NASA for NOAA. NASA makes available APT receiving station technology to anyone wishing to obtain real-time local cloud cover images. APT stations are currently located in 87 countries, many of which have made significant investments in APT.

Over the years, NASA's wide range of cooperative and reimbursable programs has benefited both the United States and the international community. Since 1962, over 40 cooperative satellites have been placed in orbit, and since 1965, more than 60 reimbursable satellite launches have been completed. In addition to satellite projects there have been over 2,000 joint ground-based and space research activities, including sounding rocket, balloon, data investigations, and space science experiments. The following is a summary of international joint satellite projects, listed in alphabetical order by country; a (C) denotes a cooperative, and an (R) a reimbursable, project.

Canada

Alouette I, September 1962

Alouette II, November 1965
Satellites for ionospheric research (C)

International Satellite for Ionospheric Studies (ISIS) I, January 1969

Isis II, March 1971 (C)

Telesat F (Anik)
November 13, 1982
The second satellite launched from the Shuttle on STS-5.

Since 1972 NASA has launched Canada's Telesat, the world's first satellite system to use geostationary satelllites for domestic telecommunications. In orbit each Telesat satellite is designated *Anik,* the Eskimo word for brother. (R)

Communications Technology Satellite (CTS)
January 1976–June 1979

The Remote Manipulator System (RMS), developed by the National Research Council of Canada, was first tested on STS-2 aboard the orbiter Columbia.

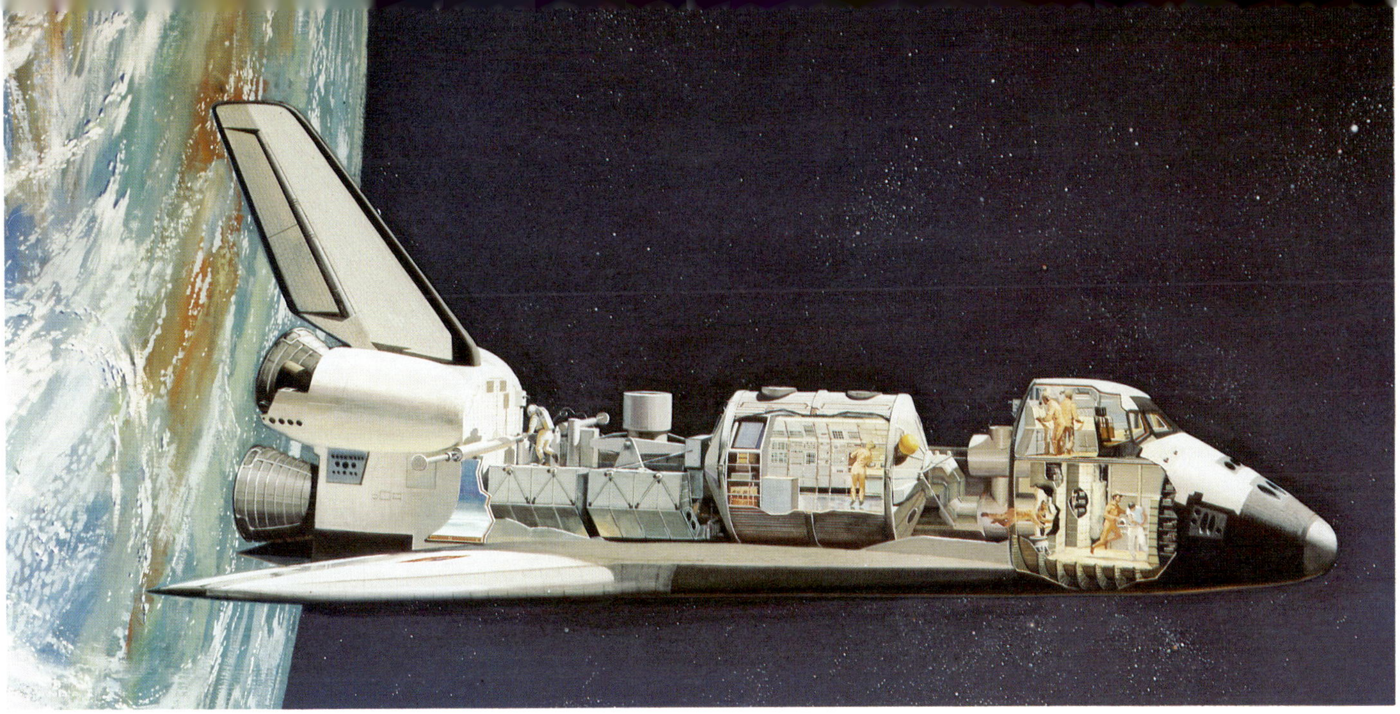

A joint US-Canadian program to advance communications via satellite; used for a number of experiments in health, education, and business. (See Chapter III) (C)

Remote Manipulator System (RMS)
November 1981–June 1983
Canada designed and built the Remote Manipulator System (RMS) for use on the Shuttle to deploy and retrieve payloads; first tested on STS-2 in November 1981, a year later NASA formally accepted the RMS, when it was declared ready for operational use. (C)

On STS-7, June 1983, the RMS's deployment/retrieval capabilities will be tested with the Shuttle Pallet Satellite (SPAS). A camera mounted on SPAS will provide photo coverage during deployment, free flight, retrieval, and reberthing on the Shuttle.

SARSAT
March 28, 1983
Canada, France, the US, and the Soviet Union are cooperating on an experimental satellite-aided search and rescue project to aid in rescue of ships and planes in distress. (See USSR) (C)

European Space Agency (ESA)

[formerly European Space Research Organization (ESRO)]
Membership: Belgium, Denmark, France, Federal Republic of Germany, Ireland, Italy, The Netherlands, Spain, Sweden, Switzerland, and the United Kingdom. Canada and Austria have "observer" status; Norway is an "Associate" member.

Technicians check ESA's Highly Eccentric Orbit Satellite (HEOS) prior to launch by NASA.

International Radiation Investigation Satellite [IRIS (ESRO/II)]
May 1968
Integrated study of solar radiation and cosmic rays. (C)

Aurorae (ESRO/I)
October 1968
Eight experiments integrated a study of high latitude energetic particles and their effects on the ionosphere. (C)

Highly Eccentric Orbit Satellite (HEOS)
HEOS-1, December 1968
HEOS-2, January 1972
Investigated interplanetary space, high altitude magnetosphere, and solar and cosmic rays. (R)

A key Shuttle payload is Spacelab (center), a multipurpose laboratory that will enable scientists to conduct experiments in the microgravity environment of space.

Boreas (ESRO/IB)
October 1969
Carried experiments designed to study ionospheric and auroral phenomena, particularly over the North Pole at night in winter. (R)

Thor-Delta (TD/1)
March 1972
Astronomy satellite carrying seven scientific experiments. (R)

ESRO/IV
November 1972
Investigated and measured several phenomena in the polar ionosphere. (R)

Cosmic Ray Satellite (COS/B)
August 1975
For study of cosmic gamma rays. (R)

Geodetic Earth Orbiting Satellite (GEOS/A)
April 1977
Designed to investigate waves and particles in the magnetosphere. (R)

GEOS/B
July 1978
Studied atmospheric radiation particles. (R)

International Sun-Earth Explorer (ISEE-2)
October 22, 1977
One of two spacecraft launched by a single rocket. With NASA's ISEE-I, was placed in an elliptical orbit to provide detailed data on Earth's immediate space environment and a variety of solar-terrestrial phenomena. In 1978, ISEE-3 was placed in a halo orbit to study the same phenomena from a different vantage point. (See Chapter VIII, Solar-Terrestrial Physics) (C)

International Ultraviolet Explorer (IUE)
January 26, 1978
A joint ESA-UK-US satellite project to study a wide range of celestial objects in one of the most important regions of the spectrum; an advanced telescope, the IUE complements ESA's TD-1 satellite and is establishing a system for observing by astronomers of all nations, an objective of the Space Telescope. (C)

Spacelab 1
Scheduled for September 1983
Spacelab, developed and built by ESA, is Europe's contribution to the NASA Space Transportation System. It consists of a cylindrical module in which both astronauts and civilian scientists, called payload scientists, will work and a series of unpressurized pallets which will support experiments requiring direct exposure to space. Carried in the cargo bay of the Shuttle orbiter, Spacelab will serve as a center for conducting scientific investigations not possible on Earth.

The first mission of Spacelab (Spacelab 1) will be a nine-day flight, a joint NASA-ESA mission during which over 70 investigations in five different scientific disciplines will be conducted. Both NASA and ESA are providing experiments for the mission. (See Epilogue, International) (C)

Space Telescope (ST)
Scheduled for Shuttle launch in 1986. (See Epilogue, International) (C)

International Solar Polar Mission (ISPM)
Scheduled for Shuttle launch in 1986. (See Epilogue, International) (C)

Ariel, the first international satellite, being checked on the weight and balance machine with its solar panels and antennas in orbit position.

France

FR-1
December 1965
For ionospheric research. (C)

Eole
August 1971
An experimental balloon satellite designed to gather meteorological data, measuring wind speeds at various altitudes. (C)

Symphonie A and B
December 1974 and August 1975
Communications satellites, a joint project with Germany to provide service to Europe, Africa, and South America. (R)

TIROS N
October 13, 1978
Centre Nationale D'Etudes Spatiales (CNES) provides data collection systems for the TIROS N advanced meteorological satellite and seven follow-on operational spacecraft. (C)

HEAO-3
September 20, 1979
France provided a heavy primary cosmic ray experiment for the third High Energy Astronomy Observatory (HEAO). (See Chapter VIII, Astronomy and Astrophysics) (C)

SARSAT
March 28, 1983
CNES provides receivers/processors for a joint US/French/Canadian search and rescue satellite-aided project. (See USSR) (C)

Germany

Azur
November 1969
Small satellite for particles and fields research. (C)

Symphonie A and B
December 1974 and August 1975
French-German communications satellites.

Helios I and 2
December 1974 and January 1976
Two deep-space probes for interplanetary and solar studies inside the orbit of Mercury. (C)

Shuttle Pallet Satellite (SPAS-01)
Scheduled for Shuttle launch (STS-7), June 18, 1983
The first payload to be deployed and retrieved by the Canadian RMS. Materials processing research experiments will be conducted while SPAS is in the Shuttle cargo bay; when these are finished, NASA will use the payload in a test of the RMS's deployment/retrieval capabilities. A camera mounted on SPAS will provide photo coverage during deployment, free flight, retrieval, and reberthing on the Shuttle. (R)

Active Magnetospheric Particle Tracer Explorers (AMPTE)
Scheduled for launch in 1984.
(See Epilogue, International) (C)

Spacelab D-1
Scheduled for launch in June 1985.
(See ESA, above and Epilogue, International) (R)

Galileo
Scheduled for Shuttle launch in 1986.
(See Epilogue, International) (C)

ROSAT
Scheduled for Shuttle launch in 1987.
(See Epilogue, International) (C)

India

Satellite Instructional Television Experiment (SITE)
August 1975–July 1976
Used the ATS-6 communications satellite for a program of educational telecasting to rural villages in India. Indian-produced programs in health, agriculture, education, and family planning were broadcast in 2,400 villages. (See Chapter III, Communications) (C)

INSAT 1B
Scheduled for Shuttle launch on STS-8, August 1983.
A geostationary satellite with telecommunications, community broadcasting, and meteorological capabilities. (R)

The San Marco 3, an Italian-built satellite, was launched by NASA from a launch platform off the coast of Kenya in the Indian Ocean.

Indonesia

Palapa B1 and B2
Scheduled for Shuttle launch, B1 on STS-7 in June 1983 and B2 in 1984.
To replace the first Palapa (A1, A2) communications satellites that have served the Indonesian archipelago, Thailand, Malaysia, and Singapore since 1976 and 1977. The new spacecraft will relay stronger signals and are expected to operate for eight years. (R)

Italy

San Marco I–III
December 1964–February 1974
Geostationary satellites for upper atmospheric research. (C)

Sirio
August 1977
Geostationary satellite designed to investigate trapped radiation flux, variation, and the primary electron energy spectrum. (R)

San Marco D/L
Scheduled for launch in July 1983.
Consists of two spacecraft equipped to study solar and meteorological phenomena. (C)

International Telecommunications Satellite Organization (Intelsat)

Intelsat is an international consortium formed in 1964, including over 90 member nations and managed by the Communications Satellite Corporation (COMSAT). A series of 24 commercial Intelsat satellites provided an international communications system from 1965 to 1982. Five more are scheduled for launch by 1985. (R)

Japan

NASA has launched three satellites for Japan, all (R).

Geostationary Meteorological Satellite (GMS)
July 1977

Communications Satellite (CS)
December 1977

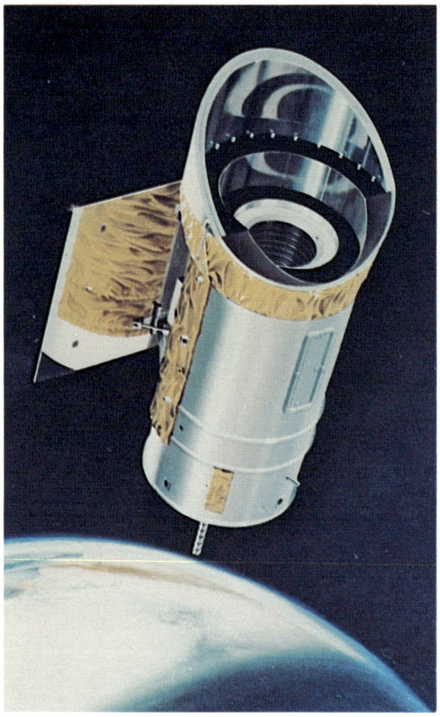

Artist's concept of the Infrared Astronomical Satellite (IRAS), a cooperative Netherlands/US/UK project.

An artist painting the US-UK identification on the Thor-Delta vehicle that launched Ariel, the first international satellite.

Soyuz spacecraft as seen from Apollo.

Broadcast Satellite Experiment (BSE)
April 1978

Netherlands

Astronomical Netherlands Satellite (ANS)
August 1974
Satellite for ultraviolet and X-ray astronomy. (C)

Infrared Astronomical Satellite (IRAS)
January 25, 1983
A cooperative Netherlands-US-UK project with a Netherlands-built spacecraft, an infrared telescope supplied by NASA, and tracking services provided by the UK; its 11-month mission is to produce an all-sky survey of discrete infrared sources. A Dutch Additional Experiment package includes three instruments, two photometers and a low-resolution spectrometer, to aid in the classification and mapping. (C)

Spain

INTASAT
November 1974
Ionospheric beacon transmitting radio signals to a world-wide network of 20 ground stations. (C)

United Kingdom

Ariel I-VI
April 26, 1962–June 2, 1979
First international satellite; the series conducted ionospheric research. (C)

Skynet 1
November 1969
Geostationary communications satellite located over the Indian Ocean. (R)

International Ultraviolet Explorer (IUE)
January 26, 1978
A joint satellite project with ESA and the US, UK provides hardware and ground support for the spacecraft and telescope, which studies the ultraviolet spectra of stars, gas clouds, planets, and comets. (C)

Infrared Astronomical Satellite (IRAS)
January 25, 1983
Cooperative project with The Netherlands and the US; the UK is providing tracking services for the spacecraft during its 11-month mission to produce an all-sky survey of discrete infrared sources. (C)

USSR

Apollo-Soyuz Test Project (ASTP)
July 15, 1975
Col. Aleksey A. Leonov
Valeriy N. Kubasov
The first joint manned space mission between the US and the USSR. (See Chapter VI, ASTP) (C)

COSPAS-SARSAT
(replaces COSPAS—still under USSR)
Cosmos 1383 (USSR), June 30, 1982
Cosmos 1447 (USSR), March 25, 1983
NOAA 5 (US), March 28, 1983
The Satellite-Aided Search and Rescue Project (COSPAS-SARSAT) is a multilateral cooperative project involving the US, Canada, France, and the USSR. US and Soviet satellites equipped with transponders are to receive emergency signals from ships and aircraft in distress and relay them to ground stations in the four countries for independent search and rescue operations. The goal of this humanitarian project is to demonstrate the effectiveness of satellites in reducing the time it takes to locate and rescue air and maritime distress victims, thus significantly increasing the possibility of saving lives. (C)

For the Classroom

1. Have your students keep a clipping file on foreign space programs, both those of individual countries and those that are cooperative ventures with the US.
2. Using the information in Chapter X, locate tracking stations around the world. How might these facilities affect the local communities?
3. Many countries have issued commemorative stamps honoring space projects. Have your students research examples and compare with US air and space commemoratives.

UNITED
STATES
UNITED

VI

Launch Vehicles

Overcoming the pull of Earth's gravity is the first challenge of any space mission. Whether small and suborbital or large and traveling to another planet, every spacecraft must be carried into space before it can do its job. NASA has a family of launch vehicles—a graduated series of multistage rockets—to accomplish its space programs.

A family of launch vehicles was developed because a number of different vehicles were required for missions that range from simple to complex. Until the Space Shuttle, launch vehicles were expendable, and the most efficient method of launching was to use a vehicle adapted to the payload, its weight, and its trajectory.

The vehicles are combinations of two or more stages, which burn one after the other, each being discarded when it is no longer needed, so only a small part of the whole vehicle is necessary to propel the spacecraft into the final orbit or space

Liftoff of the Mercury-Redstone 3, May 5, 1961.

trajectory.

When NASA was formed, its launch capability depended upon what was available and most of the vehicles were derived from the military missile program. In time, additional vehicles were developed, using both solid and liquid propellant rockets, specifically to acquire a variety of launch vehicle combinations suited to the expanding space exploration program.

From the small Scout, which is still used for small payloads, to the mighty Saturn V that took men to the Moon and was last used to put the Skylab space station into Earth orbit, NASA's launch vehicles evolved and improved both as a group and as individual vehicles. Each vehicle is constantly uprated and finally removed from service when no longer needed. In 1962 Scout could put 100 kilograms (220 lbs) and Delta, several hundred kilograms, into near-Earth orbit; Delta also could send 25 kg (55 lbs) to Mars or Venus. Ten years later Scout's performance was doubled and Delta could send 340 kg (749 lbs) to the near planets. These two vehicles remain the workhorses of the space program. First used in 1960 to put Echo into orbit, Delta is still operational, and in fact Delta was responsible for the first successful launch of 1983, the Infrared Astronomical Satellite.

NASA not only launches its own spacecraft, but it conducts many launches for commercial organizations, other Federal agencies, other nations, and multi-national groups. For such missions NASA is reimbursed for the cost of the vehicle and launch services.

NASA owns launch sites at the Eastern and Western Space and Missile Centers (ESMC and WSMC) in Florida and California and the Wallops Flight Facility in Virginia, and has access to the San Marco launch complex off the east coast of Africa owned by Italy.

The following list and accompanying chart will introduce the primary launch vehicles that NASA has used through its 25-year history.

Redstone

Adapted by NASA from an Army ballistic missile, the Redstone was used to launch Project Mercury suborbital flights.

Mercury-Redstone

1960–61, flew successfully five times after an initial failure. Two unmanned flights and one with the chimpanzee Ham preceded the first U.S. manned spaceflight by Alan B. Shephard, Jr. in May 1961, and Virgil I. Grissom's flight in July 1961.

Height: 18 m (59 ft)
25 m (83 ft) with capsule and escape tower
Single Stage
Propellant: liquid
Thrust: 35,380 kg (78,000 lbs)

Scout

Height: 20 m (68 ft)
Weight: 17,463 kg (38,500 lbs)
Payloads: 108 kg (240 lbs) (300 nm orbit)
First Stage: Algol 11B
Length: 9.07 m (30.8 ft)
Diameter: 1.14 m (4 ft)
Propellant: solid
Thrust: 39,916 kg (88,000 lbs) at sea level
Guidance: strapped down gyros (stages 1, 2, 3)
Second Stage: Castor
Length: 6.3 m (20.7 ft)
Diameter: 78.74 cm (31 in)
Propellant: solid
Thrust: 27,669 kg (61,000 lbs) vacuum
Third Stage: Antares X-259
Length: 3.5 m (11.5 ft)
Diameter: 76 cm (30 in)
Propellant: solid
Thrust: 10,432 kg (23,000 lbs) vacuum
Fourth Stage: Altair (spin stabilized)
Length: 149 cm (59 in)
Diameter: 45 cm (18 in)
Propellant: solid
Thrust: 2,630 kg (5800 lbs) vacuum
Status: Operational
Launch Pads: Wallops, WSMC, San Marco
The only NASA vehicle to use solid propellants exclusively.

Delta

Height: 35.4 m (116 ft)
Weight: 51,800 kg (114,200 lbs)
Payload: 399 kg (880 lbs) (300 nm orbit) 68 kg (150 lbs) (escape)
First Stage: Thor
Length: 17 m (55.9 ft)
Diameter: 2½ m (8 ft)
Propellant: liquid
Thrust: 77,100 kg (170,000 lbs) sea level
Verniers: (2) 453 kg (1000 lbs) each
Guidance: Radio (stages 1, 2)
Second Stage: Delta
Length: 6 m (20.6 ft)
Diameter: 1.3 m (4.3 ft)
Propellant: liquid
Thrust: 3400 kg (7500 lbs) vacuum
Third Stage: Altair (spin stabilized)
Length: 150 cm (59 in)
Diameter: 45 cm (18 in)
Propellant: solid
Thrust: 2630 kg (5800 lbs) vacuum
Status: Operational
Launch Pads: KSC, WSMC
Note: Originally called Thor-Delta and consisted of the Thor stage inherited from the Department of Defense plus the second stage of Vanguard modified somewhat and called Delta—thus the Thor-Delta original name. This was later augmented by adding Castor strap-on solid fuel motors to the Thor creating the nomenclature TAD (Thrust Augmented Delta). TAD could place a 590 kg (1300-lb) satellite into a 300-nm orbit or rocket a 113.3 kg (250-lb) spacecraft to escape velocity.

Thor-Agena D

Height: 23.3 m (76.3 ft)
Weight: classified
Payload: 725 kg (1600 lbs) (300 nm orbit)
First Stage: Thor
Length: 17 m (55.9 ft)
Diameter: 2.5 m (8 ft)
Engine: MB3 Blk 11
Propellant: liquid
Thrust: 77,100 kg (170,000 lbs) sea level
Verniers: (2) 453 kg (1000 lbs) each
Guidance: radio
Second Stage: Agena D
Length: 6.3 m (20.9 ft)
Diameter: 1.5 m (5 ft)
Engine: Bell 8096 (restartable)
Propellant: liquid
Thrust: 7,257 kg (16,000 lbs) vacuum
Guidance: inertial
Status: Operational
Launch Pads: WSMC
Note: The Thrust-Augmented Thor-Agena (TAT) has three rockets strapped to its first stage, bringing total first-stage thrust to 150,594 kg (332,000 lbs). (See note for Delta.) TAT can launch a 2200-lb satellite into a 300-nm orbit.

Atlas

The first stage Atlas launch vehicle was adapted from the first Air Force ICBM. Modified Atlases have been used in several multistage vehicles to launch both manned and unmanned missions.

Mercury-Atlas

Height: 20.5 m (67.3 ft)
29 m (95.3 ft) with capsule and escape tower

Diameter: 3 m (10 ft)
Propellant: liquid
Thrust: 139,797 kg (308,000 lbs) sea level
First used for John Glenn's orbital flight in February 1962, the Atlas launched all succeeding Project Mercury orbital flights.

Atlas-Agena D

Height: 36.6 m (120 ft)
Weight: Classified
Payloads: 2699 kg (5950 lbs) (300 nm orbit)
First Stage: Atlas D
Length: 20.5 m (67.4 ft)
Diameter: 3 m (10 ft)
Propellant: liquid
Thrust: 175,996 kg (388,000 lbs) sea level
Verniers: (2) 226.8 kg (500 lbs) each.
Guidance: radio
Second Stage: Agena D
Length: 6.3 m (20.9 ft)
Diameter: 1.5 m (5 ft)
Engine: Bell 8096 (restartable)
Propellant: liquid
Thrust: 7,257 kg (16,000 lbs) vacuum
Guidance: inertial
Status: Operational
Launch Pads: WSMC & ESMC
A versatile multi-purpose two-stage vehicle used to place unmanned spacecraft in Earth orbit or into the proper trajectory for planetary or deep-space probes. It was also used as the rendezvous target vehicle for the Gemini spacecraft in 1965–66.

Atlas-Centaur

Height: 40.8 m (134 ft)
Weight: 136,079 kg (300,000 lbs)
Payload 3856 kg (8500 lbs) (300 nm)
1043 kg (2300 lbs) (escape)
First Stage: Atlas D (modified)
Length: 23 m (75 ft)
Diameter: 3 m (10 ft)
Propellant: liquid
Thrust: 431,000 lbs sea level
1.7 million newtons
Verniers: (2) 453 kg (1000 lbs) each
Guidance: inertial (stages 1, 2)
Second Stage: Centaur
Length: 9.75 m (32 ft)
Diameter: 3 m (10 ft)
Engines: (2) RL-10 A-3
Propellant: liquid
Thrust: 13,608 kg (30,000 lbs) vacuum
133,450 newtons
Status: Operational
Launch Pads: ESMC
Centaur was the first high-energy, liquid-hydrogen liquid-oxygen propelled upper stage. Developed by NASA, it has been used in combination with Atlas and Titan boosters to launch both Earth-orbital satellites and interplanetary space probes. The Centaur can be restarted several times, which gives flexibility in launch times.

Prelaunch view of Titan-Centaur 5 with Helios B.

Titan

Titan, an Air Force ICBM, was modified by NASA as (1) Titan II for Project Gemini, 1964–66, and (2) Titan III a decade later for large payloads. Titan IIIC launched the ATS-6 communications satellite in 1974; Titan III-Centaur, the Viking and Voyager missions; and Titan III-E/Centaur, Helios 1 and 2 toward the Sun.

Titan III-E/Centaur

First launched in 1974, had an overall height of 48.8 m (160 ft). The Titan III-E booster was a two stage liquid-fueled rocket with two large solid-propellant rockets attached. At liftoff the solid rockets provided 10.7 million newtons (2.4 million lbs) of thrust.

Space Transportation System

The launch system for the Space Shuttle consists of an expendable External Tank (ET), which contains the propellants used for liftoff and ascent by the orbiter's three main engines, and two Solid Rocket Boosters (SRB's). Launched in a conventional manner, the Shuttle's Main Engines and the SRB's produce approximately 30,800,000 newtons of thrust.

At 45 kilometers (28 mi) above Earth the SRB's separate, descend by parachute, and are recovered in the ocean. Eight minutes into the flight, at approximately 110 kms (68.3 mi) altitude, the ET propellants are exhausted; the tank separates from the orbiter and disintegrates upon entry into the atmosphere.

Solid Rocket Boosters

The SRB's provide the major portion of the thrust at the time of liftoff. They are the largest solid rocket boosters ever built, the first to be used to launch humans into space, and the first designed for reuse.
Length: 45.46 m (149 ft)
Diameter: 3.70 m (12 ft)
Mass empty: 82,879 kg each (222,195 lbs)

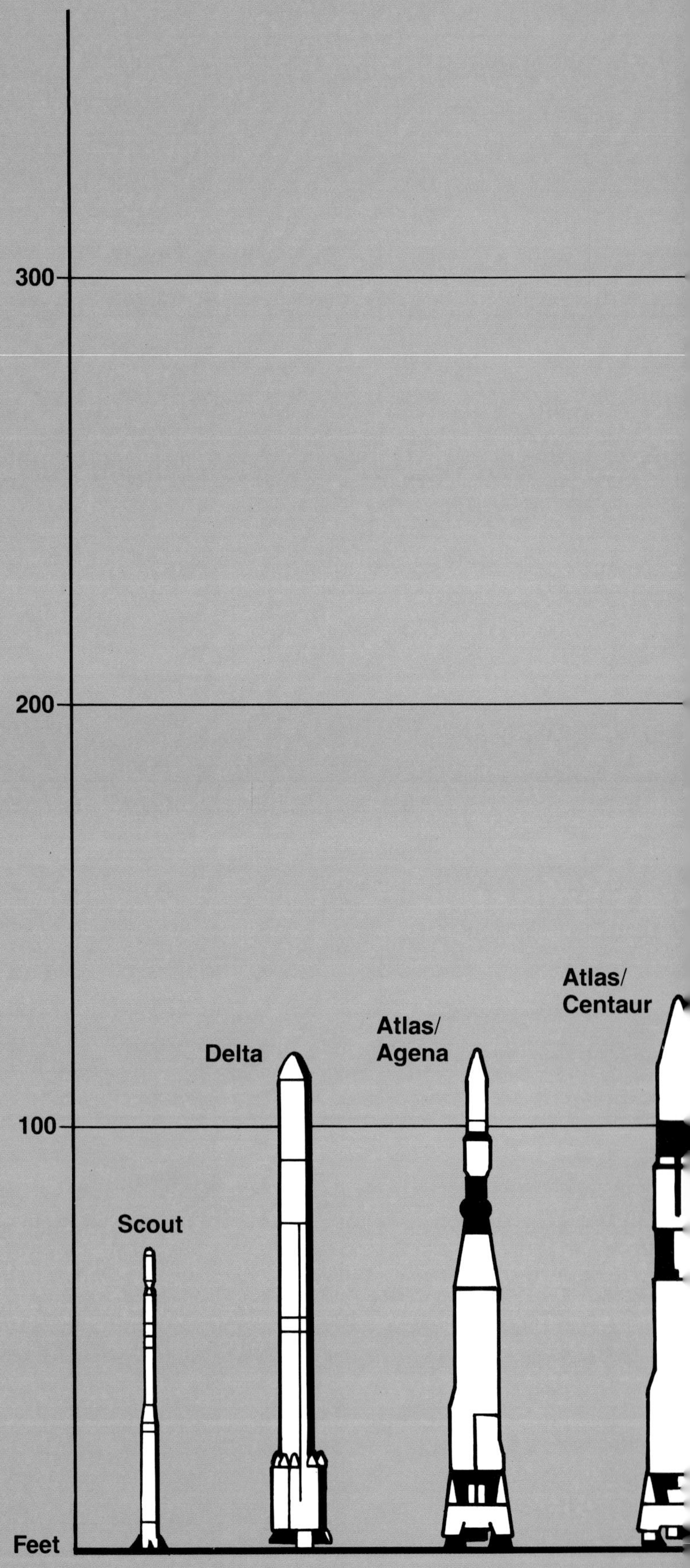

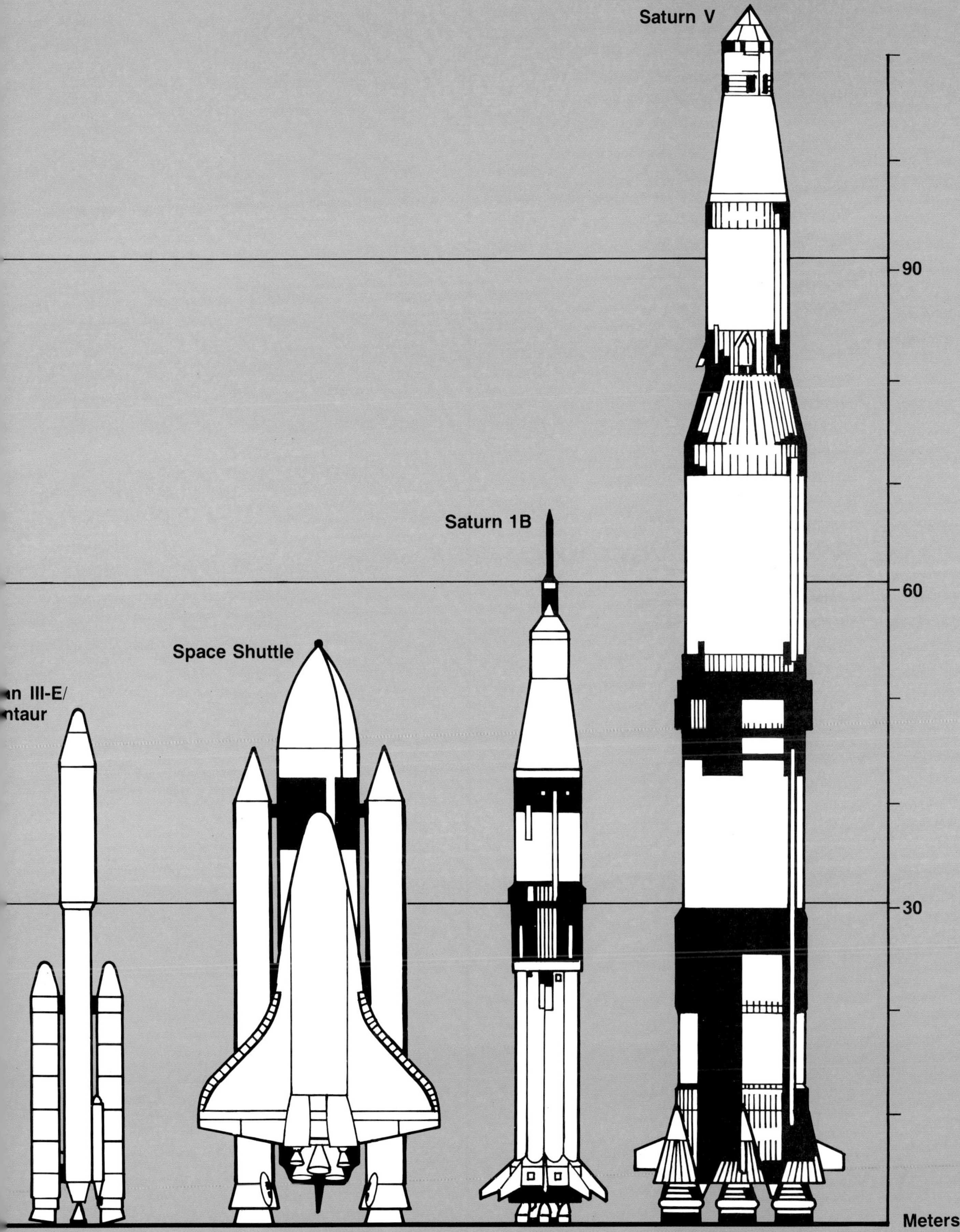

Saturn V
Saturn 1B
Space Shuttle
an III-E/
ntaur
90
60
30
Meters

Propellant mass: 503,627 kg each (1,350,206 lbs)
Thrust: 12,899,200 newtons each at sea level

External Tank
Length: 47 m (154 ft)
Diameter: 8.38 m (27.5 ft)
Mass empty: 37,452 kg (100,407 lbs)
Propellant: liquid
Propellant mass: 710,801 kg (1,905,633 lbs)

Saturn I

Height: 58 m (190 ft) with payload
Weight: 528,440 kg (1,165,000 lbs) with payload
Payloads: 10,205 kg (22,500 lbs) (100 nm orbit)
First Stage: S-1
Length: 25 m (82 ft)
Diameter: 6.6 m (21.6 ft)
Propellant: liquid
Thrust: 682,209 kg (1,504,000 lbs) sea level
Guidance: inertial (stages 1, 2)
Second Stage: S-IV
Length: 12 m (40 ft)
Diameter: 5.5 m (18 ft)
Propellant: liquid
Thrust: 40,823 kg (90,000 lbs) vacuum
Status: All flights completed successfully
Launch Pads: ESMC

Saturn I
1961–65, was a large capacity launch vehicle. Ten successful flights tested the rocket's structure, measured the performance of its motors, studied the effects of micrometeoroids on the spacecraft, and diagnosed other aspects of vehicle performance.

Uprated Saturn 1

Height: 68 m (223 ft) with payload
Weight: 589,676 kg (1,300,000 lbs)
Payloads: 18,143 kg (40,000 lbs) (100 nm orbit)
First Stage: S-IB
Length: 24 m (80 ft)
Diameter: 6.6 m (21.6 ft)
Propellant: liquid
Thrust: 725,755 kg (1,600,000 lbs) sea level, 7.1 million newtons
Guidance: inertial (stages 1, 2)
Second Stage: S-IVB
Length: 18 m (59 ft)
Diameter: 6.6 m (21.7 ft)
Propellant: liquid
Thrust: 90,719 kg (200,000 lbs) vacuum
Launch Pads: ESMC

Saturn IB
1966–75, the "uprated" Saturn, was developed to test Apollo hardware in Earth orbit. Four such tests were flown between 1966 and 1968. Saturn IB also launched Skylab 2, 3, and 4 in 1973 and the Apollo-Soyuz Test Project in 1975.

Saturn V

Height: 111 m (363 ft) with payload
Weight: 2,766,942 kg (6,100,000 lbs)
Payloads: 129,275 kg (285,000 lbs) (100 nm orbit) 43,092 kg (95,000 lbs) escape

One of the two solid rocket boosters used in the launch of STS-1 floating in its horizontal, or "log," mode for towing to Florida where it is being refurbished for reuse.

First Stage: S-IC
Length: 42 m (138 ft)
Diameter: 10 m (33 ft)
Propellant: liquid
Thrust: 3,515,377 kg (7,750,000 lbs) sea level, 34.5 million newtons
Guidance: inertial (stages 1, 2, 3)
Second Stage: S-II
Length: 24 m (80 ft)
Diameter: 10 m (33 ft)
Propellant: liquid
Thrust: 453,597 kg (1,000,000 lbs) vacuum
Third Stage: S-IVB
Length: 18 m (59 ft)
Diameter: 6.6 m (21.7 ft)
Propellant: liquid
Thrust: 90,719 kg (200,000 lbs) vacuum
Launch Pads: KSC 39A, B

Saturn V
1967–73, was the large launch vehicle developed for the Apollo lunar missions. It launched 12 successful flights, putting 27 men into lunar orbit, 12 of whom landed on the Moon. America's most powerful rocket, it carried out its last scheduled manned mission on December 7, 1972, when it launched Apollo 17. It was last used on May 14, 1973, when it lifted the unmanned Skylab space station into Earth orbit.

Saturn V with the Apollo spacecraft stood 111 m (363 ft) tall, and developed 34.5 million newtons (7.75 million lbs) of thrust at liftoff.

For the Classroom

1. Research topics:
   - The history of rocketry
   - Solid vs. liquid propellant rockets
   - New propellants that would make long-duration flights to deep space possible
2. For book reports, suggest biographies of rocket pioneers.
3. Experiment with model rocketry by purchasing commercially-produced solid rocket engines and rocket body kits (available at many hobby and toy stores and through mail order catalogs). Be sure your students follow the included instructions for constructing these rockets and check with local authorities for any regulations governing model rocketry.

   CAUTION: Constructing rocket engines, whether liquid or solid propellants are used, is a very dangerous activity when participated in by amateurs. Literally hundreds of students, teachers, and home experimenters have been seriously injured by exploding rockets. Propellant performance, chamber bursting strength, and nozzleshape are design and construction problems beyond the scope of most amateur experimenters. Model rocketry is an excellent substitution to amateur rocketry.

VII

Space Flight

"... *I believe that this Nation should commit itself to achieving the goal, before this decade is out, of landing a man on the moon and returning him safely to earth. No single space project in this period will be more exciting, or more impressive to mankind, or more important for the long-range exploration of space; and none will be so difficult or expensive to accomplish. . . . But in a very real sense, it will not be one man going to the moon—we make this judgment affirmatively—it will be an entire nation. For all of us must work to put him there."* President John F. Kennedy, Special Message to Congress, May 25, 1961

When President Kennedy delivered his national objective to Congress, the spaceflight of fiction was already becoming reality. Project Mercury had tested the spacecraft and Redstone booster (MR-1A) that the program would use; Ham, a chimpanzee named for the Holloman Aerospace Medical Center, had

The first American space walk. During the third orbit of the Gemini IV flight, Astronaut Edward H. White II floated into space secured to the spacecraft by a 25-foot umbilical line and a 23-foot tether line wrapped together with gold tape to form one cord.

been a passenger in a successful suborbital flight (MR-2); and 20 days before the President's address, the first manned suborbital flight had been accomplished.

The Congress approved and the American public enthusiastically supported the expanded and ambitious long-term space exploration program. Project Mercury with its solo pilots, Project Gemini and its two-man crews, and Project Apollo with a crew of three became a methodical progression to complete the U.S. commitment to a lunar landing.

The basic research, the program decisions, the engineering tests, the trained staff, all overlapped during the decade of these three programs. The entire country, through major contractors and thousands of subcontractors, brought the programs to fruition. Technological lessons were learned and managerial techniques refined.

Project Mercury ended in 1963. Twenty years later it seems elementary. But its pioneering experiments laid the groundwork for great achievements.

Project Gemini, in turn, provided the experience of how to live and maneuver in space. It established an experienced ground team—in the control room, at tracking stations, in industry. It readied America for the fulfillment of its national commitment.

Project Apollo, an extraordinary technological and scientific accomplishment, proclaimed the United States a leader in space exploration. It was a bridge between the early manned spaceflight programs and Skylab, the application of what they learned. It provided the technology for the international experiment, the Apollo-Soyuz Test Project.

Project Mercury

Initiated in 1958, completed in 1963, Project Mercury was the United States' first man-in-space program. The objectives of the program, which made six manned flights from 1961 to 1963, were specific:

- To orbit a manned spacecraft around Earth;
- To investigate man's ability to function in space;
- To recover both man and spacecraft safely.

All were met by the fourth flight, and new objectives of longer missions, different in quality and quantity of orbits, were added.

The Astronauts

In April 1959 seven military jet test pilots were introduced to the public as America's first astronauts: Lt. M. Scott Carpenter, USN; Capt. L. Gordon Cooper, Jr., USAF; Lt. Col. John H. Glenn, Jr., USMC; Capt. Virgil I. Grissom, USAF; Lt. Comdr. Walter M. Shirra, Jr., USN; Lt. Comdr. Alan B. Shepard, Jr., USN; and Capt. Donald K. Slayton, USAF. Six of the seven would make a Mercury flight. Slayton was grounded for medical reasons, but remained as director of the astronaut office. In 1975, returned to flight status, he served as Docking Module Pilot on the Apollo-Soyuz flight.

The Spacecraft

The first U.S. spaceship was a cone-shaped one-man capsule with a cylinder mounted on top. Two meters (6 ft, 10 in) long, 1.9 meters (6 ft, 2½ in) in diameter, a 5.8 meter (19 ft, 2 in) escape tower was fastened to the cylinder of the capsule. The blunt end was covered with an ablative heat shield to protect it against the 3000° heat of entry into the atmosphere.

The Mercury program used two

Project Mercury Astronauts, whose selection was announced only six months after NASA was formally established, included: Front row, left to right, Walter M. Schirra, Jr., Donald K. Slayton, John H. Glenn, Jr., and M. Scott Carpenter; back row, Alan B. Shepard, Jr., Virgil I. Grissom, and L. Gordon Cooper, Jr.

Astronaut Alan Shepard being lifted into the helicopter after his successful suborbital flight.

launch vehicles: A Redstone for the suborbital and an Atlas for the four orbital flights. (See Chapter V). Prior to the manned flights, unmanned tests of the booster and the capsule, carrying a chimpanzee, were made.

Each astronaut named his capsule and added the numeral 7 to denote the teamwork of the original astronauts.

The Manned Flights

Mercury-Redstone 3 (MR-3), *Freedom 7*
May 5, 1961
Alan B. Shepard, Jr.
15 minutes, 22 seconds
Suborbital flight that successfully put the first American in space.

Mercury-Redstone 4 (MR-4), *Liberty Bell 7*
July 21, 1961
Virgil I. Grissom
15 minutes, 37 seconds
Also suborbital; successful flight but the spacecraft sank shortly after splashdown.

Mercury-Atlas 6 (MA-6), *Friendship 7*
February 20, 1962
John H. Glenn, Jr.
4 hours, 55 minutes
Three-orbit flight that placed the first American into orbit.

Mercury-Atlas 7 (MA-7), *Aurora 7*
May 24, 1962
M. Scott Carpenter
4 hours, 56 minutes
Confirmed the success of MA-6 by duplicating the flight.

Mercury-Atlas 8 (MA-8), *Sigma 7*
October 3, 1962
Walter M. Schirra, Jr.
9 hours, 13 minutes
Six-orbit engineering test flight.

Mercury-Atlas 9, *Faith 7*
May 15–16, 1963
L. Gordon Cooper, Jr.
34 hours, 19 minutes
Last Mercury mission; completed 22 orbits to evaluate effects of one day in space.

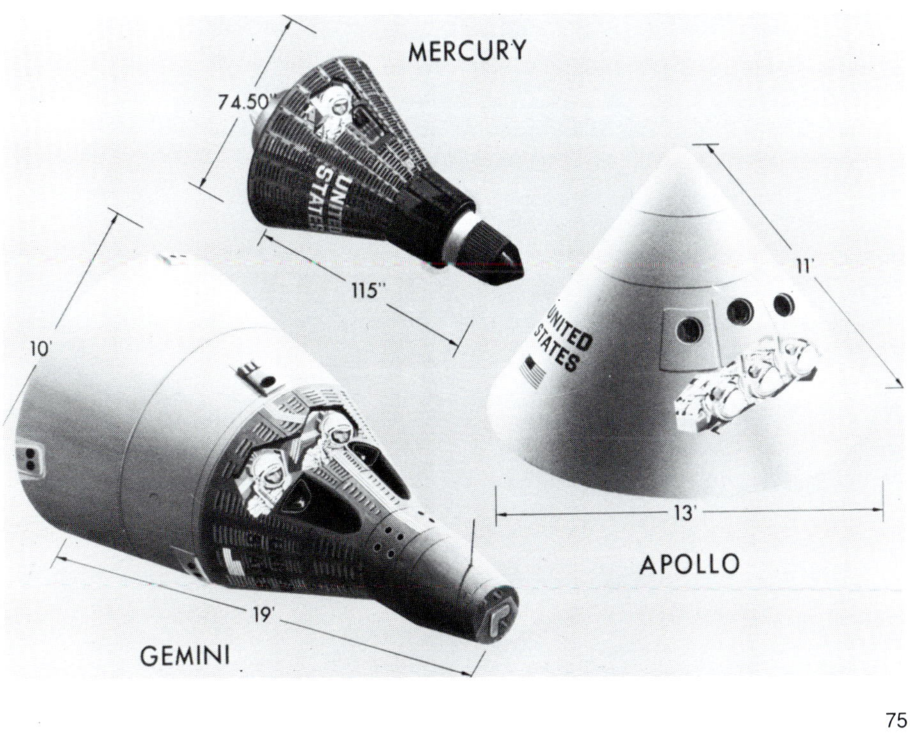

Sketches showing the comparative sizes of the Mercury capsule, the Gemini spacecraft, and the Apollo Command Module.

Project Gemini

The second U.S. manned space program was announced in January 1962. Its two-man crew gave it its name, Gemini, for the third constellation of the Zodiac and its twin stars, Castor and Pollux. Gemini involved 12 flights, including two unmanned flight tests of the equipment.

Like Mercury's, its major objectives were clear-cut:

- To subject man and equipment to space flight up to two weeks in duration;
- To rendezvous and dock with other orbiting vehicles and to maneuver the docked combination by using the target vehicle's propulsion system;
- To perfect methods of entering the atmosphere and landing at a preselected point on land. Its goals were also met, with the exception of a land landing, which was cancelled in 1964.

The Spacecraft

The spacecraft was an enlargement of the familiar Mercury capsule—5.8m (19 ft) long, 3m (10 ft) in diameter, and about 3810 kilograms (8400 pounds) in weight. Engineering changes simplified maintenance and made it more maneuverable for the pilots. The Titan II rocket, more powerful than the Redstone, placed the larger spacecraft into orbit.

Sometimes referred to as Gemini-Titan for the craft and its launch vehicle, each flight was designated by a Roman numeral. Only the first capsule was nicknamed; Command Pilot Virgil Grissom called it the *Molly Brown* in reference to his Mercury spacecraft that sank.

Suit technician assisting Gemini VI Pilot Thomas Stafford during suiting up as Command Pilot Walter Schirra looks on.

The Manned Flights

Gemini III, *Molly Brown*
March 23, 1965
4 hours, 53 minutes
Virgil I. Grissom, John W. Young
First manned Gemini flight, three orbits.

Gemini IV
June 3–7, 1965
James A. McDivitt, Edward H. White II
97 hours, 56 minutes
Included first extravehicular activity (EVA) by an American; White's "space walk" was a 22-minute EVA exercise.

Gemini V
August 21–29, 1965
L. Gordon Cooper, Jr., Charles Conrad, Jr.
8 days, 21 hours
First use of fuel cells for electrical power; evaluated guidance and navigation system for future rendezvous missions. Completed 120 orbits.

Gemini VII
December 4–18, 1965
13 days, 18 hours, 35 minutes
Frank Borman, James A. Lovell, Jr.
When the Gemini VI mission was scrubbed because its Agena target for rendezvous and docking failed, Gemini VII was used for the rendezvous instead. Primary objective was to determine whether humans could live in space for 14 days.

Gemini VI
December 15–16, 1965
Walter M. Shirra, Jr., Thomas P. Stafford
25 hours, 51 minutes
First space rendezvous accomplished with Gemini VII, station-keeping for over five hours at distances from 0.3 to 90 m (1 to 295 ft).

Gemini VIII
March 16, 1966
Neil A. Armstrong, David R. Scott
10 hours, 41 minutes
Accomplished first docking with another space vehicle, an unmanned Agena stage. A malfunction caused uncontrollable spinning of the craft; the crew undocked and effected the first emergency landing of a manned U.S. space mission.

Gemini IX
June 3–6, 1966
Thomas P. Stafford, Eugene A. Cernan
3 days, 21 hours
Rescheduled from May to rendezvous and dock with augmented target docking adapter (ATDA) after original Agena target vehicle failed to orbit. ATDA shroud did not completely separate, making docking impossible. Three different types of rendezvous, two hours of EVA, and 44 orbits were completed.

Gemini X
July 18–21, 1966
John W. Young, Michael Collins
5 days
First use of Agena target vehicle's propulsion systems. Spacecraft also rendezvoused with Gemini VIII target vehicle. Collins had 49 minutes of EVA standing in the hatch and 39 minutes of EVA to retrieve experiment from Agena stage. 43 orbits completed.

Photograph of the Gemini VII spacecraft taken through the hatch window of Gemini VI during rendezvous and station-keeping maneuvers at an altitude of approximately 160 miles on December 15, 1965 (above).

"Angry Alligator" was the Gemini IX crew's description of the Augmented Target Docking Adaptor with its shroud partly open and still attached (upper right).

Gemini XI
September 12–15, 1966
Charles Conrad, Jr., Richard F. Gordon, Jr.
5 days, 8 hours
Gemini record altitude, 1,189.3 km (739.2 mi) reached using Agena propulsion system after first orbit rendezvous and docking. Gordon made 33-minute EVA and two-hour standup EVA. 44 orbits.

Gemini XII
November 11–15, 1966
James A. Lovell, Jr., Edwin E. Aldrin, Jr.
3 days, 22 hours, 34 minutes
Final Gemini flight. Rendezvoused and docked with its target Agena and kept station with it during EVA. Aldrin set an EVA record of 5 hours, 30 minutes for one space walk and two stand-up exercises.

Project Apollo

"That's one small step for a man, one giant leap for mankind." The national effort that enabled Astronaut Neil Armstrong to speak those words as he stepped onto the lunar surface, fulfilled a dream as old as humanity. But Project Apollo's goals went beyond landing Americans on the Moon and returning them safely to Earth:

- To establish the technology to meet other national interests in space;
- To achieve preeminence in space for the United States;
- To carry out a program of scientific explorations of the Moon; and
- To develop man's capability to work in the lunar environment.

The cumulative experience of Mercury and Gemini started Apollo with confidence. The mighty Saturn launch vehicles—for both Earth orbit and lunar flights—had perfect test flights.

Apollo 204/Apollo 1

January 27, 1967. Tragedy struck on the launch pad during a preflight test for Apollo 204, the first Apollo manned mission. Astronauts Virgil Grissom, Edward White, and Roger Chafee lost their lives when a fire swept through the Command Module. Had it flown, the mission would have been Apollo 1, a designation that was officially assigned to it.

The investigation and re-engineering of the spacecraft based on the findings caused an 18-month delay. But in the fall of 1968 Apollo was ready for flight.

The Spacecraft

Apollo was a three-part spacecraft: the command module (CM), the crew's quarters and flight control section; the service module (SM) for the propulsion and spacecraft support systems (when together, the two modules are called CSM); and the lunar module (LM), to take two of the crew to the lunar surface, support them on the Moon, and return them to the CSM in lunar orbit.

The flight mode, lunar orbit rendezvous, was selected in 1962. The boosters for the program were the Saturn IB for Earth orbit flights and the Saturn V for lunar flights. The crews that made lunar flights where both CM and LM were involved, again selected call names. In the list of flights, crews are named in the following order: Commander, CM Pilot, LM Pilot. The call names for the spacecraft, in parentheses, are the CM and LM, respectively.

The Manned Flights

Apollo 7
October 11–12, 1968
Walter M. Schirra, Jr.
Donn F. Eisele
R. Walter Cunningham
10 days, 20 hours
163 Earth orbits. First manned CSM operations in lunar landing program. First live TV from manned spacecraft. Rendezvous with upper stage.

Apollo 8
December 21–27, 1968
Frank Borman
James A. Lovell, Jr.
William A. Anders
6 days, 3 hours
In lunar orbit 20 hours, with 10 orbits. First manned lunar orbital mission. Support facilities tested. Photographs taken of Earth and Moon. Live TV broadcasts.

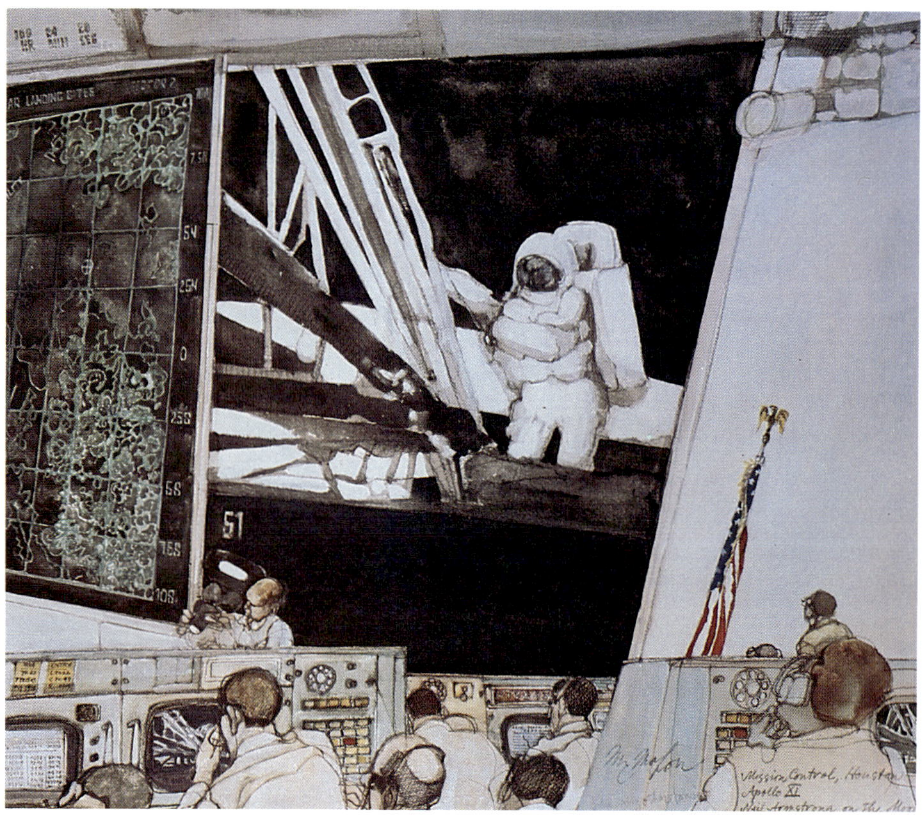

***The Eagle Has Landed* by Franklin McMahon**

A composite lunar scene showing scientist-astronaut Harrison Schmitt standing next to a boulder at the Apollo 17 Taurus-Littrow landing site and the lunar rover in the distance.

Apollo 9 (Gumdrop and Spider)
March 3–13, 1969
James A. McDivitt
David R. Scott
Russell L. Schweickart
10 days, 1 hour
First manned flight of all lunar hardware in Earth orbit. Schweickart performed 37 minutes EVA. Human reactions to space and weightlessness tested in 152 orbits. First manned flight of lunar module.

Apollo 10 (Charlie Brown and Snoopy)
May 18–26, 1969
Eugene A. Cernan
John W. Young
Thomas P. Stafford
8 days, 3 minutes
Dress rehearsal for Moon landing. First manned CSM/LM operations in cislunar and lunar environment; simulation of first lunar landing profile. In lunar orbit 61.6 hours, with 31 orbits. LM taken to within 15,243 m (50,000 ft) of lunar surface. First live color TV from space. LM ascent stage jettisoned in orbit.

Apollo 11 (Columbia and Eagle)
July 16–24, 1969
Neil A. Armstrong
Michael Collins
Edwin E. Aldrin, Jr.
8 days, 3 hours, 18 minutes
First manned lunar landing mission and lunar surface EVA. *"Houston, Tranquillity Base here. The Eagle has landed."*—July 20, Sea of Tranquillity.

1 EVA of 2 hours, 31 minutes. Flag and instruments deployed; unveiled plaque on the LM descent stage with inscription: "Here Men From Planet Earth First Set Foot Upon the Moon. July 1969 A.D. We Came In Peace For All Mankind." Lunar surface stay time, 21.6 hours; 59.5 hours in lunar orbit, with 30 orbits. LM ascent stage left in lunar orbit. 20kg (44 lbs) of material gathered.

Apollo 12 (Yankee Clipper and Intrepid)
November 14–24, 1969
Charles Conrad, Jr.
Richard F. Gordon, Jr.
Alan L. Bean
10 days, 4 hours, 36 minutes
Landing site: Ocean of Storms. Retrieved parts of the unmanned Surveyor 3, which had landed on the Moon in April 1967. Apollo Lunar Surface Experiments Package (ALSEP) deployed. Lunar surface stay-time, 31.5 hours; in lunar orbit 89 hours, with 45 orbits. LM descent stage impacted on Moon. 34kg (75 lbs) of material gathered.

Apollo 17 view of Earth.

Apollo 13 (Odyssey and Aquarius)
April 11–17, 1970
James A. Lovell, Jr.
John L. Swigert, Jr.
Fred W. Haise, Jr.
5 days, 22.9 hours
Third lunar landing attempt. Mission aborted after rupture of service module oxygen tank. Classed as "successful failure" because of experience in rescuing crew. Spent upper stage successfully impacted on the Moon.

Apollo 14 (Kitty Hawk and Antares)
January 31–February 9, 1971
Alan B. Shepard, Jr.
Stuart A. Roosa
Edgar D. Mitchell
9 days
Landing site: Fra Mauro. ALSEP and other instruments deployed. Lunar surface stay-time, 33.5 hours; 67 hours in lunar orbit, with 34 orbits. 2 EVAs of 9 hours, 25 minutes. Third stage impacted on Moon. 42 kg (94 lbs) of material gathered, using hand cart for first time to transport rocks.

Astronaut Edwin E. Aldrin, Jr., lunar module pilot, descending the steps of Lunar Module ladder as he prepares to walk on the Moon.

Apollo 15 (Endeavor and Falcon)
July 26–August 7, 1971
David R. Scott
James B. Irwin
Alfred M. Worden
12 days, 17 hours, 12 minutes
Landing site: Hadley-Apennine region near Apennine Mountains. 3 EVAs of 10 hours, 36 minutes. Worden performed 38 minutes EVA on way back to Earth. First to carry orbital sensors in service module of CSM. ALSEP deployed and scientific studies begun on Apollo 11, 13, and 14 missions continued; scientific payload landed on Moon doubled. Improved spacesuits gave increased mobility and stay-time. Lunar surface stay-time, 66.9 hours. Lunar Roving Vehicle (LRV), electric-powered, 4-wheel drive car, traversed total 27.9 km (17 mi). In lunar orbit 145 hours, with 74 orbits. Small sub-satellite left in lunar orbit for first time. 6.6 kgs (169 lbs) of material gathered.

Apollo 16 (Casper and Orion)
April 16–27, 1972
John W. Young
Thomas K. Mattingly II
Charles M. Duke, Jr.
11 days, 1 hour, 51 minutes
Landing site: Descartes Highlands. First study of highlands area. Selected surface experiments deployed, ultraviolet camera/spectrograph used for first time on Moon, and LRV used for second time. Lunar surface stay-time, 71 hours; in lunar orbit 126 hours, with 64 orbits. Mattingly performed 1 hour in-flight EVA. 95.8 kg (213 lbs) of lunar samples collected.

Apollo 17 (America and Challenger)
December 7–19, 1972
Eugene A. Cernan
Ronald B. Evans
Harrison H. Schmitt
12 days, 13 hours, 52 minutes
Last lunar landing mission. Landing site: Taurus-Littrow, highlands and valley area. 3 EVAs of 22 hours, 4 minutes. Evans performed trans-Earth EVA lasting 1 hour, 6 minutes. First scientist-astronaut to land on Moon, Schmitt. Sixth automated research station set up. LRV traverse total 30.5 km. Lunar surface stay-time, 75 hours. In lunar orbit 17 hours. 110.4 kg, (243 lbs) of material gathered.

Apollo 17 spacecraft floating to splashdown.

Skylab

America's first experimental space station. Designed for long duration missions, Skylab program objectives were twofold: To prove that humans could live and work in space for extended periods, and to expand our knowledge of solar astronomy well beyond Earth-based observations. Successful in all respects despite early mechanical difficulties, three three-man crews occupied the Skylab workshop for a total of 171 days, 13 hours. It was the site of nearly 300 scientific and technical experiments: medical experiments on humans' adaptability to zero gravity, solar observations, and detailed Earth resources experiments. Skylab remained in orbit for more than six years, completing 34,981 orbits. It returned to Earth July 11, 1979 and disintegrated in the atmosphere, scattering debris over the Indian Ocean and the sparsely settled region of Western Australia.

The Spacecraft

Skylab was a modification of the original Apollo equipment. Elongated through the use of a "dry" third stage of the Saturn V rocket, the spacecraft was completely outfitted as a workshop area before launch.

The Flights

Skylab 1
May 14, 1973
Unmanned
The station was launched into orbit by a Saturn V booster. Almost immediately, technical problems developed due to vibrations during lift-off. A critical meteoroid shield ripped off taking one of the craft's two solar panels with it; a piece of the shield wrapped around the other panel keeping it from deploying.

Skylab was maneuvered so its Apollo Telescope Mount (ATM) solar panels faced the Sun to provide as much electricity as possible. Because of the loss of the meteoroid shield, however, this positioning caused workshop temperatures to rise to 52° Celsius (126° F).

Skylab 2
May 25–June 22, 1973
Charles Conrad, Jr.
Paul J. Weitz
Joseph P. Kerwin
28 days, 50 minutes
First manned mission. The crew rendezvoused with Skylab on the fifth orbit; after making substantial repairs, including deployment of a parasol sunshade which cooled the inside temperatures to 23.8°C (75°F) by June 4, the workshop was in full operation. In orbit the crew conducted solar astronomy and Earth resources experiments, medical studies, and five student experiments; 404 orbits and 392 experiment hours were completed; three EVAs totalled five hours, 34 minutes.

Skylab 3
July 28–September 25, 1973
Alan L. Bean
Jack R. Lousma
Owen K. Garriott

Astronaut Thomas Stafford and Cosmonaut Aleksey Leonov in the Soyuz orbital module.

59 days, 11 hours
Continued maintenance of the space station and extensive scientific and medical experiments. Completed 858 Earth orbits and 1,081 hours of solar and Earth experiments; three EVAs totalled 13 hours, 42 minutes.

Skylab 4
November 16, 1973–February 8, 1974
Gerald P. Carr
William R. Pogue
Edward G. Gibson
84 days, 1 hour
Last of the Skylab missions; included observation of the Comet Kohoutek among numerous experiments. Completed 1,214 Earth orbits and four EVAs totalling 22 hours, 25 minutes.

View of the Skylab space station cluster in Earth orbit taken from the Skylab 3 Command/Service Module during the "fly around" inspection prior to docking.

Apollo-Soyuz Test Project

Apollo-Soyuz was the first international manned spaceflight. It was designed to test the compatibility of rendezvous and docking systems for American and Soviet spacecraft, to open the way for international space rescue as well as future joint manned flights.

The Spacecraft

The existing American Apollo and Soviet Soyuz spacecraft were used. The Apollo spacecraft was nearly identical to the one that orbited the Moon and later carried astronauts to Skylab. The Soyuz craft was the primary Soviet spacecraft used for manned flight since its introduction in 1967. A docking module was designed and constructed by NASA to serve as an airlock and transfer corridor between the two craft.

The Flight

ASTP
July 15–24, 1975
Thomas P. Stafford
Vance D. Brand
Donald K. Slayton
9 days, 7 hours, 28 minutes
The Soyuz was launched just over seven hours prior to the launch of the Apollo CSM. Apollo then maneuvered to rendezvous and docking 52 hours after the Soyuz launch. The Apollo and Soyuz crews conducted a variety of experiments over a two-day period. After separation, Apollo remained in space an additional 6 days. Soyuz returned to Earth approximately 43 hours after separation.

Space Transportation System (STS)

The Space Transportation System includes the first reusable spacecraft. The Space Shuttle is a four-part vehicle: a reusable orbiter, which resembles a jetliner, mounted piggyback on an expendable liquid propellant external tank (ET) and two recoverable and reusable solid rocket boosters (SRBs).

Designed for routine use of space and to provide the first commercial space operations, the Shuttle operates in low Earth orbit. In space, it is a base to deploy payloads; it will also be used to repair and service satellites in orbit, to retrieve satellites, and to serve as a platform for scientific research.

Because of its versatility and reusability, the Shuttle is expected to reduce the cost of space operations significantly.

The Spacecraft

The manned component of the STS is the orbiter. Capable of being used up to 100 times, it is a 68,000-kg (150,000-lb) craft measuring about 38 m (125 ft) in length with a wingspan of about 24 m (78 ft). Its payload bay is 18.3 m (60 ft) long and 4.6 m (15 ft) in diameter.

It is launched in the conventional manner, and, in orbit, operates like a spacecraft. When returning to Earth, upon entry into the atmosphere, the orbiter sails back like a glider and lands at a designated ground location.

Assembly of the first Space Shuttle vehicle was completed in November 1980 with the mating of the orbiter Columbia to its external tank and solid rocket boosters.

Touchdown of the Columbia at the conclusion of STS-1.

The Flights

Approach and Landing Tests

In September 1976, the Space Shuttle Orbiter 101, the Enterprise (named for the *Star Trek* spaceship), was rolled out from the Rockwell International final assembly facility. The following January, it was transported overland by a 90-wheel trailer to Dryden Flight Research Facility. There, during 1977, it underwent a series of 13 Approach and Landing Tests (ALT) from atop a modified Boeing 747 Shuttle Carrier Aircraft (SCA), to verify its aerodynamic and flight control characteristics, avionics, structures, and mechanical systems performance.

Phase I
February–March
Five flights of the mated SCA/Enterprise. The orbiter was unmanned and its systems turned off.

Phase II
Three manned captive flights of the Enterprise/SCA with astronauts at the controls of the orbiter verified crew procedures and systems operations.

Two crews of two astronauts were selected for the ALT flights: Fred W. Haise, Jr. and C. Gordon Fullerton and Joe H. Engle and Richard H. Truly.

Flights 1 (June 18) and 3 (July 26) were flown by Haise and Fullerton; Flight 2 (June 28) by Engle and Truly.

Phase III
During five flights the orbiter was released from the SCA and glided to a runway landing at Edwards, California. Flights 1 (August 12), 3 (September 23), and 5 (October 26) were flown by Haise and Fullerton. Flights 2 (September 13), and 4 (October 12) were flown by Engle and Truly.

Orbital Flights

Selected from sea vessels used in world exploration, the names of the first four orbiting Space Shuttle craft are Columbia, Challenger, Discovery, and Atlantis.

In the list of the flights, the commander's name is placed first, then the pilot's and mission specialists'.

STS-1, Columbia
April 12–14, 1981
C: John W. Young
P: Robert L. Crippen
2 days, 6 hours
First flight of the reusable Space Transportation System; proved the feasibility of ground landings and module reuse.

STS-2, Columbia
November 12–14, 1981
C: Joe H. Engle
P: Richard H. Truly
2 days, 6 hours
Second flight tested the craft's Remote Manipulator System (RMS) (See Chapter IV, Canada) and other components of the vehicle; cut short by the failure of a fuel cell.

STS-3, Columbia
March 22–30, 1982
C: Jack R. Lousma
P: C. Gordon Fullerton
8 days
First Shuttle mission to be launched on the originally scheduled day, the major goal was thermal testing of the orbiter and continuation of experiments with Shuttle systems.

STS-4, Columbia
June 27–July 4, 1982
C: Thomas K. Mattingly II
P: Henry W. Hartsfield, Jr.
7 days, 1 hour
Final test flight. Although the two solid rocket boosters sank in the Atlantic Ocean, the mission was otherwise routine, and the Space Shuttle was declared operational.

The President and Mrs. Reagan inspect the Columbia after its STS-4 landing at Edwards, California, on July 4, 1982. Left to right: Astronauts Henry Hartsfield, pilot, and Thomas Mattingly, commander; Mrs. Reagan; President Reagan. Columbia's fourth flight completed the test phase of the Space Transportation System.

Prime crew for the first orbital flight test of the Space Shuttle Columbia: John W. Young, left, commander, and Robert L. Crippen, pilot.

STS-5, Columbia
November 11–16, 1982
C: Vance D. Brand
P: Robert F. Overmyer
MS: Dr. Joseph P. Allen
MS: Dr. William B. Lenoir
5 days, 2 hours
First operational flight and first with mission specialists aboard, who were trained in satellite deployment, EVA, and use of the RMS. Two communications satellites, Satellite Business Systems' SBS-3 and Telesat Canada's Anik C-3, were deployed from the cargo bay.

STS-6, Challenger
April 4–9, 1983
First flight of the Challenger. Originally scheduled for launch in January 1983, liftoff was delayed due to a variety of technical problems. In orbit, the mission began placement of the Tracking and Delta Relay Satellite System (TDRSS). The system will provide high-capacity communications between numerous spacecraft and Earth.

STS-7, Challenger
Scheduled launch, June 18, 1983
C: Robert L. Crippen
P: Frederick H. Hauck
MS: Dr. Sally K. Ride
MS: John M. Fabian
MS: Dr. Norman E. Thagard
The 6-day flight crew includes the first American woman in space. Payloads: a German shuttle pallet satellite (SPAS), which will be used in a test of the Canadian RMS deployment/retrieval capabilities; the second Office of Space and Terrestrial Applications package (OSTA-2); and Canadian and Indonesian communications satellites.

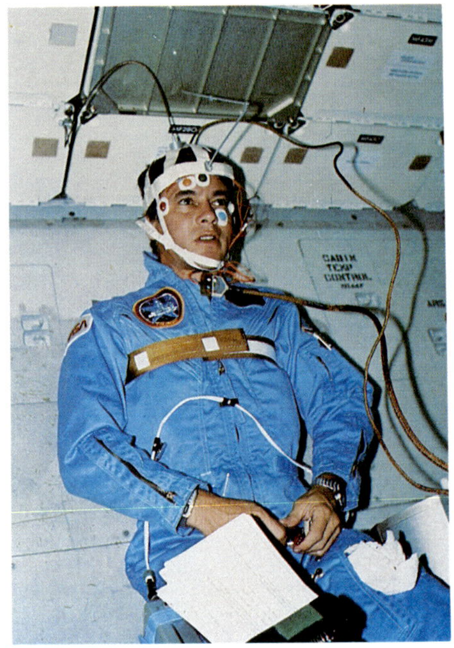

Astronaut Joseph Allen participating in a biomedical test in the middeck area of the Columbia during STS-5. Electrodes connected to his face monitored his responses in zero gravity.

STS-7, the first five-member crew: Ride, Crippen,Hauck (front row, left to right); Fabian (left) and Thagard, (right).

STS-8 crew: (left to right) Daniel C. Brandenstein, Dale A. Gardner, Richard H. Truly, William E. Thornton, and Guion S. Bluford, Jr.

STS-8, Challenger
Scheduled launch, August 1983
C: Richard H. Truly
P: Daniel C. Brandenstein
MS: Dale A. Gardner
MS: Guion S. Bluford, Jr.
MS: Dr. William E. Thornton
This 3-day mission will put into orbit a communications satellite for India's Department of Science.

STS-9, Columbia
Scheduled launch, Fall 1983
C: John W. Young
P: Brewster H. Shaw, Jr.
MS: Dr. Owen K. Garriott
MS: Dr. Robert A. Parker
PS: Ulf Merbold, ESA
PS: Dr. Byron K. Lichtenberg, NASA
Columbia will return to space carrying Spacelab, a reusable laboratory designed for use in missions lasting from seven to 30 days. It will be carried in the orbiter's cargo bay and allow experimenters to work in a shirtsleeve environment or on external instrument platforms exposed to space. The Spacelab crews are Payload Specialists.

SBS-3 just moments following its release from a spin table during its deployment from the cargo bay of the Columbia on STS-5.

OSTA-1, the pallet of experiments that was the Shuttle's first payload on STS-2, being lowered into the Columbia's payload bay.

STS-10, Challenger
Scheduled launch, November 1983
C: Thomas K. Mattingly II
P: Loren J. Shriver
MS: Ellison S. Onizuka
MS: James F. Buchli
A Department of Defense mission.

The Experiments

The science and engineering experiments flown aboard early Space Shuttle flights suggest the great potential of the orbiter as a platform for scientific and applications research and the benefits of routine access to low Earth orbit. The payloads have been NASA-sponsored, Getaway Special canisters, and cooperative international projects; they span the scientific disciplines; and they include experiments from the private sector, commercial concerns, students, and foreign governments.

OSTA-1

NASA's Office of Space and Terrestrial Applications (OSTA) sponsored the first scientific and applications payload (OSTA-1), which flew on STS-2. The payload was a set of instruments that involved remote sensing of land resources, atmospheric phenomena, and ocean conditions: Shuttle Imaging Radar, Shuttle Multispectral Infrared Radiometer, Feature Identification and Location Experiment, Ocean Color Experiment, Measurement of Air Pollution from Satellites, Night/Day Optical Survey of Lightning, and Heflex Bioengineering Test.

OSS-1

The first Office of Space Science experiments package (OSS-1) flew on STS-3. This group of experiments included eight mounted on a pallet in the orbiter's cargo bay—six to study the interaction of the orbiter with its environment, trap meteoroids, and monitor the buildup of contaminants in the cargo bay, and two solar experiments—and a ninth in the crew cabin, a miniature terrarium called Plant Growth Unit, to study the effects of weightlessness on plants.

Also included were an Electrophoresis Equipment Verification Test and a Monodisperse Latex Reactor, a materials-processing experiment, which were flown on later flights as well.

Shuttle Student Involvement Project (SSIP)

The SSIP is a competition to give secondary school students opportunities to develop payload experiments suitable for flight aboard the Shuttle. Each winning student has a NASA scientist and corporate sponsor who give advice and instructions on readying experiments for flight; payload assignments are given to Shuttle flights as the experiments are ready and payload space is available.

Six SSIP projects have flown. The first to fly, *Insects in Flight Motion Study,* was on STS-3. Two more flew on STS-4 and three were on STS-5.

Getaway Special

Getaway Specials, Small Self-contained Payloads, are low-cost experiments flown in canisters. Scheduled for flight on a space-available basis, they are available to educational organizations, industry, individuals, and governments. There are no stringent requirements, but the canister must meet safety criteria and the experiment must have a scientific or technological objective.

A test canister was flown on STS-3. The first operational Getaway Special flew on STS-4, a 61 × 91 cm (24 × 35 in) canister that held nine experiments designed by Utah State University students. The second flew on STS-5; STS-6 had three; and STS-7 will carry seven, including an ant colony experiment developed by Camden (NJ) high school students.

For the Classroom

1. Research topics:
   - Spaceflight in fact and fiction
   - Environmental elements that must be considered in designing space suits, capsules, laboratories, stations
   - Necessities for comfortable living on long-duration spaceflights (arrangement of living space, decor, music, books)
   - Space foods and nutritious menus
   - History of space suit design
2. For book reports, suggest autobiographies and biographies of the astronauts.
3. Have your students suggest names for future Shuttle orbiters and explain their selections.
4. Compare the Shuttle space suit with those worn by comic book characters.
5. Collect information for a class file on the Shuttle flights, the Shuttle payloads, the crews.
6. What kinds of jobs can robots do better than humans?
7. High school teachers interested in involving their students in future Shuttle Student Involvement Project competitions should write to the following address:
   Shuttle Student
   Involvement Project
   National Science Teachers
   Association
   1742 Connecticut Avenue, NW.
   Washington, DC 20009

VIII

Space Science

In 1958, when NASA received responsibility for developing space science, the Moon, the planets, the Sun, the universe seemed remote and inaccessible. Oniy Explorer 1's discovery of the Van Allen radiation belts suggested the knowledge that was to come. In the ensuing 25 years more has been learned about the universe than during all history.

NASA's space science programs began with the Pioneer spacecraft. In 1958 and 1959 eight were launched to study Earth-Moon relationships. Failures outnumbered successes at first. Then remarkable strides were made with a variety of projects—lunar and solar investigations, planetary programs, observations of distant stars and galaxies. Scientists searched for life on another planet, observed the sky with telescopes above the atmosphere, studied human performance in space—the accumulated data will occupy them for years. And the solar system became a neighborhood, the universe a

Saturn and its satellites, Tethys (outer left), Enceladus (inner left), and Mimas (right of rings), photographed by Voyager 1 on October 30, 1980, from 18 million kilometers (11 million miles).

source of boundless energy and mysterious objects.

In NASA's 25th year, an event that will occur only once in human history will take place when Pioneer 10 becomes the first spacecraft to leave the solar system, June 13, 1983, 5:00 a.m. PDT.

Astronomy and Astrophysics

Two major areas of space science that NASA has studied intensively are astronomy and astrophysics. Satellites and space probes have investigated planetary atmospheres, including Earth's; radio physics; interplanetary space; and wavelengths of the electromagnetic spectrum. The results have provided extensive information about Earth, its relation to the rest of the solar system, our galaxy, and the universe beyond.

The astronomy and astrophysics research has been especially valuable because the investigations were conducted beyond Earth's atmosphere which distorts images seen through telescopes and filters out radiation wavelengths. Scientists have been able to observe clearly and study closely phenomena they are unable to view from Earth.

In the list that follows, the first date noted for each project is the launch date.

Explorer

The Explorers comprise a long series of small scientific satellites that vary widely in design and purpose. In more than 50 missions they have studied Earth's shape and surface, near-Earth space and interplanetary space, and astronomical and astrophysical phenomena. Many had project names that were used before they orbited, but were supplanted by Explorer designations once in orbit.

Explorer 42, Small Astronomy Satellite (SAS-1)
December 12, 1970
First of NASA's Small Astronomy Satellites and first orbiting X-ray satellite; mapped the universe in X-ray wavelengths for four years and discovered X-ray pulsars and evidence of black holes.

First US satellite to be launched by another country, it was launched from the San Marco platform off the coast of Kenya on that country's Independence Day, and was christened Uhuru, Swahili for "Freedom."

Explorer 48 (SAS-2)
November 1972
Performed sky survey of high energy gamma radiation, discovered gamma ray pulses from the pulsar in the Vela X supernova remnant, and the first radio pulsar in a binary system.

Explorer 53 (SAS-3)
May 1975–April 1979
Was a prolific producer of X-ray astronomy data long past its design lifetime of one year.

SAS-3's detailed studies of the positions, spectra, and time variations of individual X-ray sources strengthened the expectation that many more extremely distant extragalactic objects would be discovered by the more sensitive detectors aboard the HEAO (High Energy Astronomy Observatory) spacecraft. A main accomplishment was the discovery of several X-ray bursters, stars that emit gigantic, brief bursts of X-rays once every few hours; all the known bursters are at least a billion times farther away than the Sun.

Pegasus

Named for the winged horse of mythology, these satellites featured huge wing-like panels 30 meters (96 ft) tip to tip to determine the rate of meteoroid penetrations. Results: Interplanetary dust particles are about 10,000 times less abundant than had been indicated by earlier space experiments.

The dates noted are launch and data collection termination.

Pegasus 1
February 16, 1965–January 13, 1968

Pegasus 2
May 25, 1965–March 14, 1968

Pegasus 3
July 30, 1965–August 29, 1968

Orbiting Astronomical Observatory (OAO)

OAO-1
April 1966
First of three successfully launched Orbiting Astronomical Observatories, large sophisticated satellites for studying stars. The spacecraft was lost due to power failure and no scientific results were obtained.

OAO-2
December 1968–February 1973
Took the first ultraviolet photographs of stars; discovered hydrogen clouds around comets detectable only in ultraviolet light.

OAO-3 (Copernicus)
August 1972
Named Copernicus after launch as part of the international celebration of the 500th anniversary of the astronomer's birth. Carried first instrumentation for ultraviolet spectroscopy of stars from Earth orbit. Results: Found evi-

Drawing of Pegasus being deployed and wings unfolding.

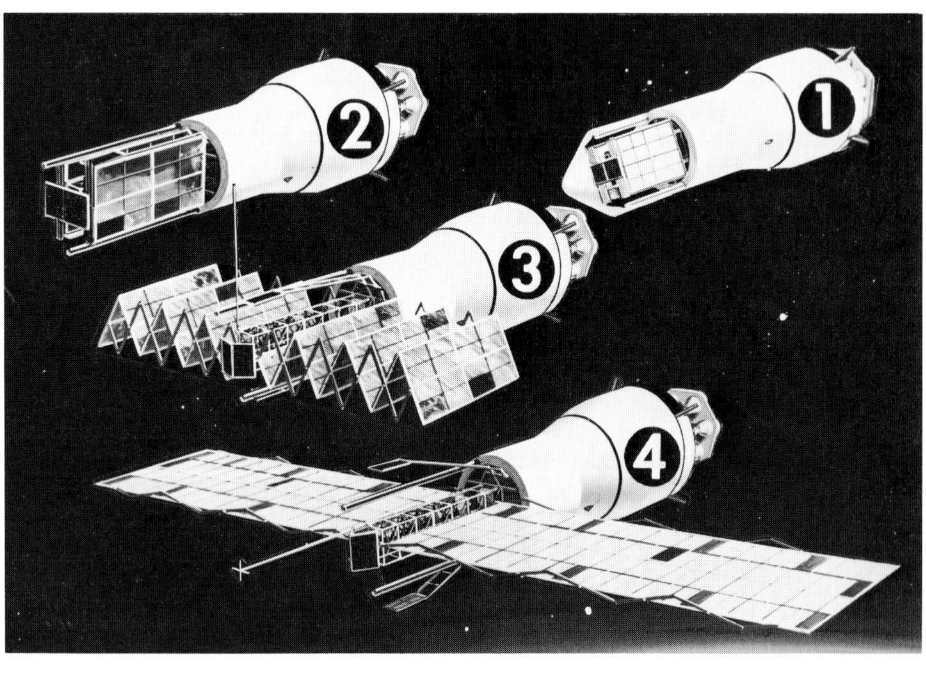

dence for "Open Universe" theory, which says expansion of universe will never end; bulk of neutral interstellar gas is contained in small dense clouds rather than uniformly distributed; a theory that low-density cavities in interstellar space are caused by supernova explosions.

High Energy Astronomy Observatory (HEAO)

A three-mission program of large observatories designed to study high energy rays—X-ray, gamma rays, and cosmic rays—that cannot be studied through ground-based telescopes because of Earth's atmosphere. HEAO-1 and HEAO-3 were scanning missions, HEAO-2 a pointing mission.

HEAO-1
August 1977–March 1979
Conducted a general X-ray sky survey. The quality of data was so excellent that its design life of six months was extended to 17.

Three important results: The catalogue of celestial X-ray sources was increased from 350 to nearly 1500; a new black hole candidate was located near the Constellation Scorpius, bringing the total to four; and a universal hot plasma constituting a major fraction of the mass of the universe was discovered, as well as a cloud of dust and gas with a mass equal to a million billion Suns enveloping a supercluster of galaxies.

HEAO-2
November 1978–June 1981
A pointing mission to point to specific X-ray sources identified by HEAO-1. Popularly known as the Einstein Observatory, it carried the largest X-ray telescope built to that time to search for information about quasars, pulsars, exploding galaxies, and black holes.

Discoveries: Double the number of known X-ray sources; new quasars and a new class of stars known as O stars; new clues to some of the most intriguing phenomena in the universe—remnants of supernovae, pulsars, neutron stars, "cosmic boosters" in globular clusters; images of X-ray sources in galaxies outside our own; oldest, most distant clusters of galaxies yet observed.

HEAO-3
September 1979–August 1981
Conducted an all-sky survey searching for cosmic ray particles and gamma ray emissions. Provided valuable new information on the high energy processes of cosmic rays and detected strong gamma ray emission from major features along the galactic plane.

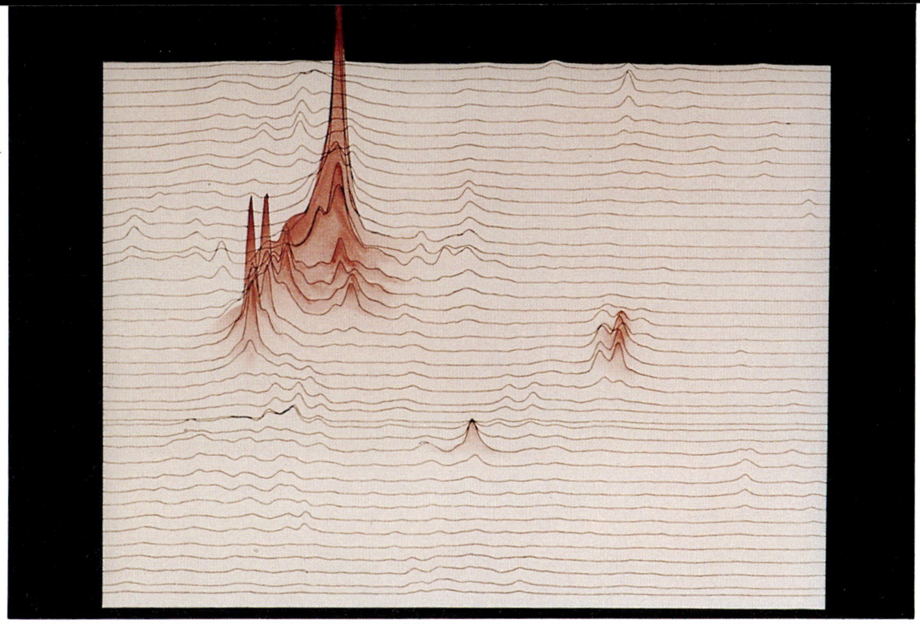

A mosaic map composed from data relayed by the Infrared Astronomical Satellite (IRAS) shows part of the Large Magellanic Cloud, the nearest galaxy outside our own Milky Way, at 155,000 light years away and at a wavelength of 100 microns.

X-ray picture taken in 1979 by the High Energy Astronomy Observatory 2 (HEAO-2) reveals a newly-discovered object (upper left) which appeared to be the most distant quasar observed. Its Red Shift of Z-2.6 indicated that the light reaching us began its journey more than 10 billion years ago.

Gravity Probe A

June 1976

A scientific probe to test Einstein's theory of relativity. Proved the dilation of time effect.

International Ultraviolet Explorer (IUE)

January 1978

An advanced telescope, it complements and extends observations made by OAO-2 and Copernicus and ESA's TD-1 satellite.

Findings: Information about the way in which galaxies are able to emit so much energy that they are receding from us at 90% of the velocity of light; the first discovery of the element sulfur in a comet; acetylene in the atmosphere of Saturn; a corona of 100,000°C around the Milky Way galaxy and coronae around the Magellanic Clouds.

Infrared Astronomical Satellite (IRAS)

January 25, 1983

In a near-polar orbit, IRAS is conducting the first infrared survey of the sky. An international project with a telescope supplied by NASA, a spacecraft built by The Netherlands, and tracking services supplied by the UK, it is expected to complete two sky surveys during its 11-month operational life. Initial discoveries: Two regions where stars like our Sun are being born and two comets, IRAS and IRAS-Araki-Alcock.

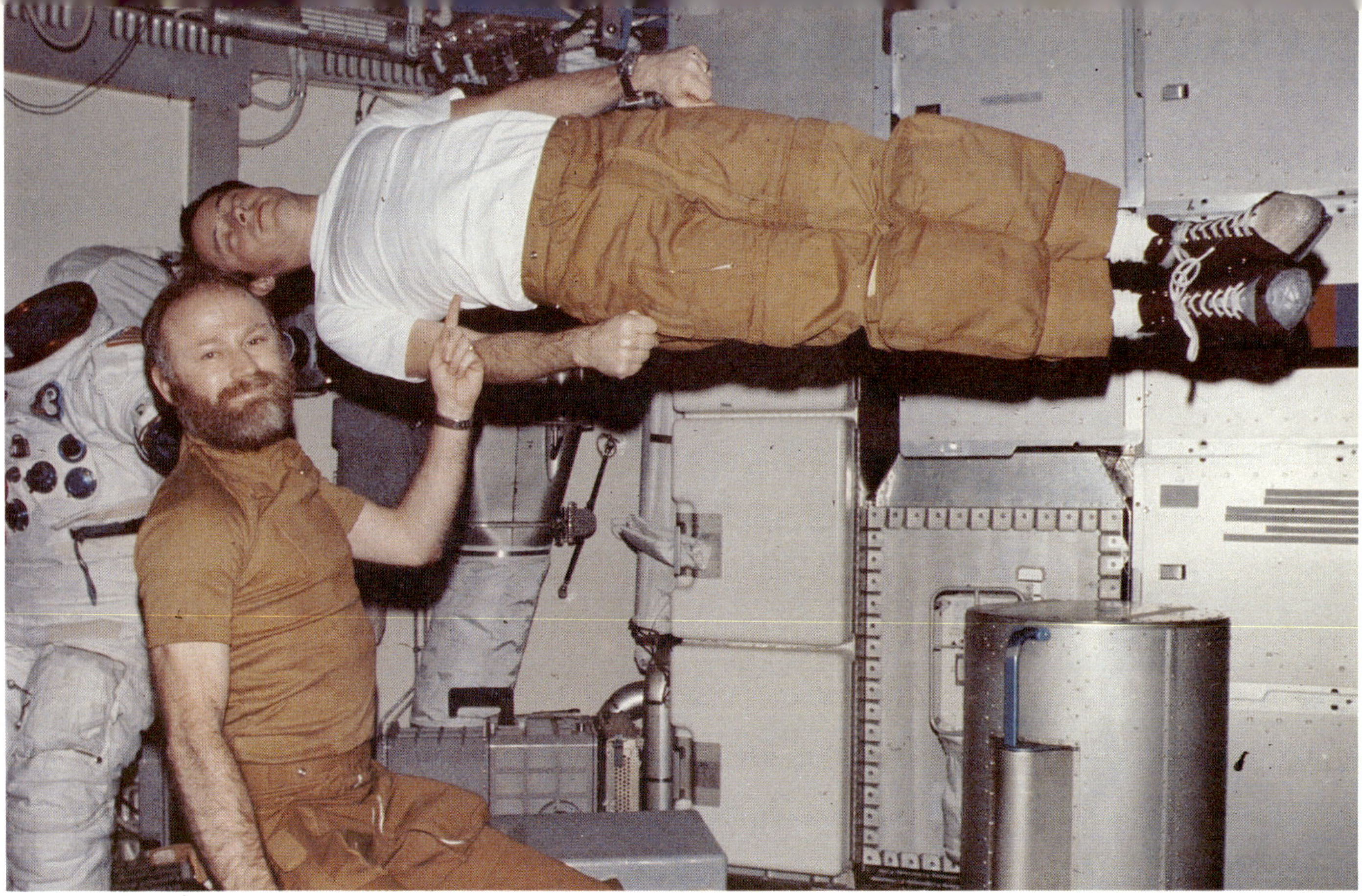

Life Sciences

The objective of life science research is to determine the requirements for human safety in space; to learn as much as possible about the function of biological, biochemical, and physiological systems in the space environment; to utilize space conditions for isolation, purification and synthesis of biological materials; to assure that Earth applications for space technology and hardware are fully exploited; and to understand the origin and distribution of life in the universe.

NASA space life sciences research has focused on studies of cellular and molecular biology, radiobiology, botany, zoology, cardiovascular system, respiration, metabolism and human performance, chemical evolution studies, and the search for life on Mars.

Biosatellite

Biosatellite 1
December 1966
First satellite to conduct biological experiments; the flight failed on recovery and only engineering tests were completed because the biological experiments were dependent upon recovery for data acquisition.

Biosatellite 2
September 7–9, 1967
Carried 13 experiments; obtained information on the effects of radiation and weightlessness on plants, cells, and low-order animal life forms such as gnats, bacteria, and larvae. Capsule recovered by aircatch.

Biosatellite 3
June 29–July 7, 1969
Biological experiments with a monkey as subject provided information on the effects of prolonged weightlessness.

Orbiting Frog Otilith (OFO-1)

November 1970–May 1971

A biological experiment designed to study the adaptability of the otolith (frog's inner ear balance mechanism) to sustained weightlessness to provide information for manned spaceflight.

Skylab

May 1973–February 1974

Skylab was designed to determine the safety of long term spaceflight for humans. During three of the Skylab missions, scientist-astronauts conducted medical experiments associated with the extension of manned spaceflight: cardiovascular science, cell biology, endocrinology, immunology, and radiobiology.

Demonstrating weightlessness in Skylab.

Apollo-Soyuz Test Project (ASTP)

July 1975

Many ASTP experiments were conducted by the US alone or by the US and USSR jointly. Joint science projects included experiments with zone-forming fungi and a microbial exchange test. US experiments included microorganism growth (of protea vulgaris), fish embryonic development, and genetic experiments (in seeds). In two medical experiments, an electric field was used to separate blood samples into their constituents to help determine whether the unique space environment might offer a better means of isolating viruses, enzymes, and other small particles for analysis.

Solar-Terrestrial Physics

Studies of Sun-Earth relationships, the interplanetary medium, and astronomical studies of the Sun began in 1958 with a variety of spacecraft. The broad spectrum of investigations included: air density studies over the entire globe to measure how density variations are affected by latitude, season, and local solar time; Earth's atmosphere in regions of high solar energy absorption to measure temperature, composition, and density; energetic particles experiments; ionization in the immediate vicinity of Earth to determine the nature, dynamic behavior, and distribution of charged particles, electrons, and ions; Earth's magnetosphere to measure how this area affects Earth's weather and climate; and solar physics studies.

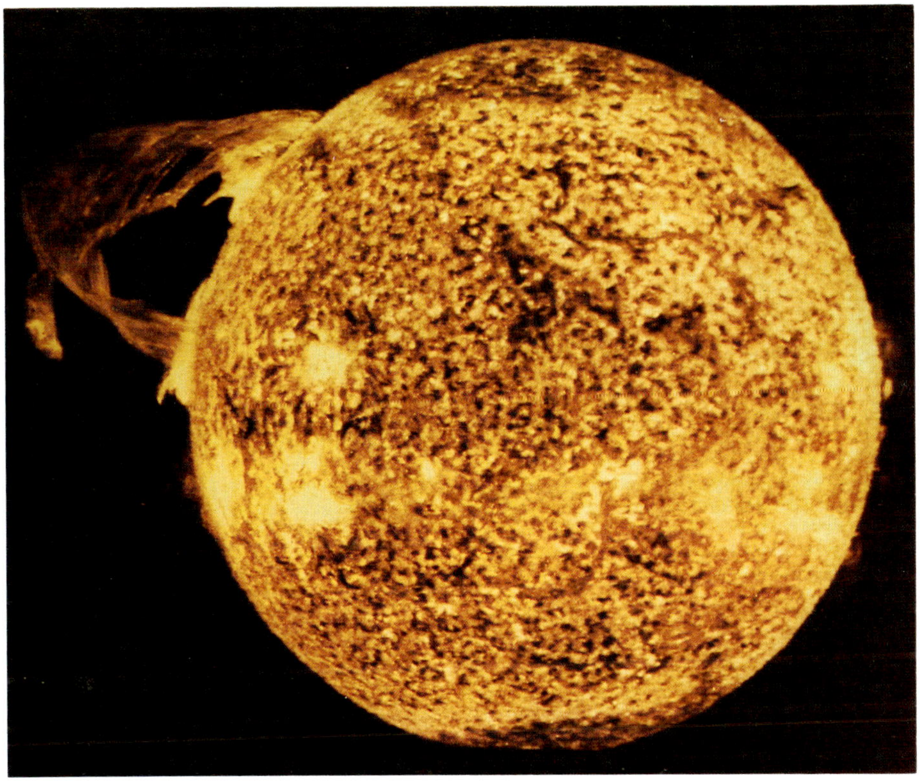

This photograph of the Sun, taken by Skylab 4, shows one of the most spectacular solar flares ever recorded.

Beacon

In October 1958, August 1959, and March 1964 Beacons 1, 2, and 3 were launched for ionospheric research using a small radio beacon. Launch vehicles for all three failed to place them in orbit.

Explorer

Explorer 8
November 1960
To study the ionosphere and atmospheric composition; confirmed the existence of a helium layer in the upper atmosphere.

Explorer 9
February 1961–April 1964
First in a series of Air Density Explorers designed to determine the effect of thin air on satellite motion; the spacecraft was a 3.7-m (12-ft) diameter inflatable sphere covered with aluminum foil over Mylar.

Explorer 12
August 1961–September 1963
First of four Energetic Particle Explorers designed to study injection, trapping, and loss mechanisms of Earth's trapped radiation belts; identified the Van Allen Belt as a magnetosphere.

Explorers 18, 21, 28, 33–35, 41, 43, 47, 50 (Interplanetary Monitoring Platforms)
The Interplanetary Monitoring Platform (IMP) series, numbering ten, was launched from 1963 to 1973. The majority were placed in highly elliptical Earth orbits and one in lunar orbit, to study interplanetary radiation and magnetic fields in cislunar space (between Earth and Moon), and to extend knowledge of Sun-Earth-Moon relationships.

Results: Advance warning of solar flare activity that might affect astronauts in space; first accurate measurement of the interplanetary magnetic field; first mapping of the shock front boundary of the magnetosphere and magnetopause; first detailed information on magnetosphere tail region; first evidence of energetic electrons in the neutral area which may be the source of radiation causing the aurora; and first on-line interplanetary cosmic ray monitor.

Explorer 18 (IMP-1)
November 1963–December 1965
First accurate measurement of the interplanetary magnetic field.

Explorer 34 (IMP-5)
May 1967–May 1969
Discovered that Saturn emits radio waves, as do Earth and Jupiter.

Explorer 35 (IMP-6)
July 1967
First IMP to achieve lunar orbit; discovered that positive ions from the solar wind crash directly onto the lunar surface and that a solar wind void exists directly behind the Moon.

Explorers 47 and 50 (IMP-8 and 10)
September 22, 1972 and October 25, 1973
In near-circular Earth orbit, worked in conjunction throughout a year. The two craft were frequently 180 degrees apart permitting simultaneous studies of solar phenomena from opposite sides of the Earth.

Explorer 17
April 1963
First of five Atmosphere Explorers (AE). Data on temperature, composition, density, and pressure permitted the study of global atmospheric physics; discovered a belt of neutral helium atoms around Earth.

Explorer 19
December 1963
An inflatable Air Density Explorer in polar orbit; measured the upper thermosphere and lower exosphere over the entire globe.

Explorer 20
August 1964
An Ionosphere Explorer, a topside sounder

that examined the ionosphere from above.

Explorer 22
October 1964
Measured the total electron content of the ionosphere by acting as a radio beacon.

Explorer 24
November 1964–October 1968
A 3.6 m (12-ft) sphere Air Density Explorer, half of NASA's first dual payload launch with Explorer 25.

Explorer 25 (Injun 4)
First of the Injun Explorers. Interdisciplinary project to investigate the magnetosphere (25) and correlate atmospheric density (24) information with energy measurements.

Explorer 27
April 1965
Last of the Ionosphere Explorers; obtained descriptions of Earth's gravitational field.

Explorer 30 (Solar Explorer I)
November 1965
Monitored solar X-rays for correlation with optical and radio ground-based observations during the International Quiet Sun Year.

Explorer 31 (International Satellite for Ionospheric Studies, ISIS-X)
November 1965
A cooperative project with Canada, the dual launch consisted of two spacecraft, the Canadian Alouette II and the American Direct Measurements Explorer, for ionospheric and solar research.

Explorer 32
May 1966
An atmosphere Explorer, similar to Explorer 17 but with solar cells to extend its life; for atmospheric and ionospheric research.

Explorer 37 (Solar Explorer 2)
March 1968
Joint Naval Research Laboratory (NRL)-NASA spacecraft.

Explorers 39 and 40
[Air Density (39) and Injun 5 (40)]
August 1968
Interdisciplinary project to continue detailed scientific study of density and radiation characteristics of Earth's upper atmosphere at a time of high solar activity.

Explorer 44 (SOLRAD 10)
July 1971
NRL spacecraft to monitor solar X-ray radiation and ultraviolet emissions and to improve prediction techniques of solar activity and ionospheric disturbances.

Explorer 45
November 1971
Investigated the ring-current and magnetic storms in the teardrop-shaped envelope that surrounds the Earth and mitigates the effects of particles emitted by the sun.

Explorer 52 (Hawkeye 1)
June 1974
Last of the Injun series, renamed Hawkeye for the University of Iowa-built spacecraft.

The Orbiting Solar Observatory (OSO-5) flight spacecraft undergoes checks in preparation for launch.

Launched into a polar orbit to study the solar wind/magnetic field interactions in the vicinity of the magnetic neutral point over Earth's North Pole cap.

Explorer 54
October 1975
To study atmospheric physics; although turned off after four months of operation, returned a full set of data on the daylight side of Earth's atmosphere from North to South Poles.

Explorer 55
November 1975
Last of the Atmosphere Explorers, this satellite is still investigating the chemical processes and energy transfer mechanisms that control Earth's atmosphere.

Solar Mesophere Explorer (SME)
October 1981
Designed to determine what changes occur in the ozone distribution in the upper atmosphere as a result of changes in incoming radiation.

Pioneer

Pioneer 5
March 1960
Studied the area between Earth and Sun. Provided the first data on solar flares and established the existence of interplanetary magnetic fields.

Pioneer 6–9
December 1965–November 1968
A network of solar weather satellites which circle the Sun. Measurements of the Sun and interplanetary space by the four spacecraft from widely separated points are used to predict solar storms for some 1,000 users. All four have provided data for years beyond their six-month design lives.

Orbiting Solar Observatory

The Orbiting Solar Observatory (OSO) spacecraft were the first of the observatory-class satellites and were designed for continuous observations of the Sun and its atmosphere during most of its 11-year cycle in X-ray, ultraviolet, and infrared wavelengths. Some celestial objects were studied as well.

The spacecraft consisted of a spinning "wheel" and a despun "sail" section which allowed some experiments to remain constantly pointed at the Sun.

Results: The first X-ray from a spacecraft of a beginning solar flare and of solar "streamers" (structures in the corona); first observation of the corona in white light and extreme ultraviolet; solar flare temperatures of 30 million degrees and of the Sun's poles of one million degrees Celsius.

OSO-1
March 1962
Provided 77 days of solar observations including 140 solar flares; the first extended studies of the solar ultraviolet spectrum and of solar X-ray emissions, mapped the sky in gamma radiation wavelengths, and discovered coronal holes.

OSO-2
February 1965
Scanned the solar disc to measure X-ray and ultraviolet emissions and the white light in the Sun's corona.

OSO-3
March 1967
Expanded typical OSO experiments to include

cosmic gamma rays and extreme ultraviolet radiation from the Sun.

OSO-4
October 1967
Designed to concentrate studies on shorter wavelengths and sunspot areas rather than the entire solar disc.

OSO-5
January 1969
Contained an ultraviolet spectroheliograph to map the solar disc.

OSO-6
August 1969
Expanded OSO-5 studies.

OSO-7
September 1971
Conducted experiments in conjunction with the Apollo Telescope Mount carried on Skylab. Obtained first identification of gamma ray emission lines in solar flares and first photographs of rapidly moving structures in the Sun's white light corona.

OSO-8
September 1978
Obtained the most accurate observations of the solar chromosphere and transition region; found that the number of X-ray bright points on the Sun during minimal solar activity was greater than in mid-solar cycle.

Orbiting Geophysical Observatory

The Orbiting Geophysical Observatory (OGO) missions were designed to conduct varied physical experiments within Earth's atmosphere and magnetosphere and in cislunar space (between Earth and Moon) to study Earth-Sun relationships and the Earth itself. The spacecraft, a 0.9x0.9x1.8-m (3x3x6-ft) rectangular prism with 13 appendages, demonstrated the feasibility of three-axis stabilized observatories whose design could be used repeatedly. The names EGO and POGO were developed to apply to OGO satellites in particular orbits: highly eccentric (EGO) for OGO-1, 3, and 5; and polar (POGO) for OGO-2, 4, and 6.

OGO-1
September 1964
Designed for diversified and interrelated physical experiments within Earth's atmosphere. Its orbit provided a capability for mapping and studying Earth's magnetosphere and interplanetary space. Not all the booms deployed properly and OGO-1 did not stabilize.

OGO-2
October 1965
Placed in a near-polar low-altitude orbit; carried 20 experiments to investigate the atmosphere, ionosphere, magnetic field, radiation belts, cosmic rays, micrometeorites, and solar emissions.

Artist's concept shows the OGO in orbit around Earth.

OGO-3
June 1966
Essentially the same experiments as OGO-1; remained spin-stabilized for 46 days.

OGO-4
July 1967
Obtained data during increased solar activity to complement near-solar minimum data of OGO-2.

OGO-5
March 1968
Considered most successful OGO for its number of successful experiments (22 of 25) and number of experimental hours of data (636,000).

OGO-6
June 1969
Carried 26 experiments to study interrelationship of Earth and Sun during a period of increased solar activity. Power and equipment degradation left 14 experiments operating normally, three partially operating and nine off by September 1970.

International Radiation Investigation Satellite (IRIS-1)

May 1968–May 1971
Measured radiation from the Sun and cosmic rays including X-rays, solar and Van Allen belt protons, cosmic ray protons, and high energy electrons.

Skylab

May 1973–February 1974
Scientist astronauts observed a wide range of solar conditions over a nine month period, using the Skylab Apollo Telescope Mount (ATM). The ATM carried an X-ray spectroheliograph, an X-ray telescope, a white light coronagraph, an ultraviolet spectrograph, an extreme ultraviolet spectroheliograph, and an extreme ultraviolet spectrometer-spectroheliometer.

Over 180,000 frames were taken with the ATM. In addition to the Sun, the ATM and a far-ultraviolet camera were used to view Comet Kohoutek in 1973.

Helios

Helios 1, December 1974

Helios 2, January 1976
Joint two-spacecraft project with West Germany to investigate interplanetary space and explore the near-solar region. Named for the ancient Greek god of the Sun, the probes carried instruments closer to the Sun than any previous spacecraft—within the outer edge of the corona 45 million kilometers (28 million miles) from the Sun.

International Sun-Earth Explorer (ISEE)

ISEE-1 and ISEE-2
October 22, 1977
ISEE-1, built and operated by NASA, was followed into orbit on the same day by ISEE-2, built and operated by the European Space

Agency (ESA). The craft were placed in the same elliptical orbit, but at a distance ranging from several hundred to thousands of kilometers apart, to study solar-terrestrial relationships at the outermost boundaries of Earth's magnetosphere.

ISEE-3
August 12, 1978
Placed in a giant "halo orbit" around the L-1 "Liberation Point" located between Earth and Sun about 1.6 million km (1 million miles) from Earth. The first satellite to orbit a point rather than a body, ISEE-3 measured solar wind and other phenomena while ISEE-1 and 2 measured the effect of the same phenomena on the near-Earth environment.

In 1982, it was moved to a new position to conduct an exploratory survey of Earth's magnetotail; in 1985 it will be the first to fly through a comet's tail.

Solar Maximum Mission (SMM)

February 14, 1980
Planned to coincide with the solar maximum period, the peak of the 11-year sunspot cycle, to provide scientists with observations of solar flares over a wide band of wavelengths from visible light to the gamma ray region of the spectrum.

Solar System Exploration

From its beginning NASA has been interested in the scientific study of the Moon and planets. In 1961 it had the added responsibility of securing lunar information that would be needed for a manned expedition to the Moon.

The Moon

Pioneer

Pioneer was chosen as the name for the first U.S. space probe, a series initiated for the International Geophysical Year by the Department of Defense. Like the Explorer series, NASA inherited the responsibility for the probe and kept its name.

The first eight were designed as lunar probes. Because of launch vehicle failure, the program met with little success—only Pioneer 4 reached the Moon—but Pioneer 1 discovered the radial extent of the radiation belts, Pioneer 2 returned data about the atmosphere, and Pioneer 3 discovered the second radiation belt.

Pioneer 4
March 1959
Measured particles and fields in a flyby of the Moon; entered a heliocentric orbit.

Ranger

The Ranger program was a probe series to transmit close-up black and white photographs of the Moon before crashing into the lunar surface. Three of the nine Rangers (7, 8, 9) were successful. Rangers 1 through 5 experienced technical problems which affected the success of the missions: the launch vehicles malfunctioned for Rangers 1 and 2, Rangers 3 and 5 missed the Moon, and Ranger 4 landed on the back side of the Moon and returned no data.

Ranger 7, 8, and 9 findings: A gently rolling terrain with no sharp relief; and a layer of powdery rubble, with rocks and craters down to at least one meter in diameter everywhere.

Ranger 6
January 30–February 2, 1964
Lunar impact point of hard landing within 32 km (20 mi) of target. TV system failed to operate.

Ranger 7
July 28–31, 1964
First successful Ranger mission. 4,316 high resolution TV pictures of the lunar surface were returned, with objects less than .9m (3 ft) discernible. Impact on Sea of Clouds 13–16 km (8–10 mi) from aim point. Flight time: 68 hours, 36 minutes.

Ranger 8
February 17–20, 1965
7,137 pictures returned from Sea of Tranquillity. Flight time: 64 hours, 63 minutes.

Ranger 9
March 21–24, 1965
5,814 pictures of Crater Alphonsus and vicinity returned. 4.8 km (3 mi) from target. Flight time: 64 hours, 31 minutes.

Model of the Ranger VI spacecraft.

Surveyor

Following the Ranger hardlandings, from 1966 to 1968, the Surveyor series was conducted to softland unmanned spacecraft on the Moon, survey it with TV cameras, and analyze the chemical composition of the lunar surface.

Five of the seven Surveyors were successful. They operated on the lunar surface over a combined time of 17 months, transmitted more than 17,000 pictures, and made analyses of surface and subsurface samples.

Surveyor 1
May 30–June 2, 1966
Successful soft landing in Ocean of Storms. 11,237 pictures returned; found that the surface is firm and capable of supporting machines and astronauts.

Surveyor 3
April 17–20, 1967
Landed in the Sea of Clouds and returned 6,315 pictures. First soil scoop. Piece of the spacecraft brought back by Apollo 14.

Surveyor 5
September 8–10, 1967
Soft landing in the Sea of Tranquillity. Returned over 19,000 pictures. First alpha scatter instrument analyzed chemical composition and found that the surface of the maria resembles that of terrestrial basalt lava.

Surveyor 6
November 7–10, 1967
Soft landing in the Central Bay region. Returned 30,065 pictures. First lift-off from lunar surface moved it ten feet to new location.

Surveyor 7
January 7–10, 1968
Successful soft landing on ejecta blanket adjacent to Crater Tycho. First combination of

the three major experiments: TV, alpha scatter, and surface sampler. Found that the highlands composition differs from that of the maria and is aluminum-rich.

Lunar Orbiter

Final unmanned lunar program. The Orbiters worked in conjunction with Surveyor to acquire photographic and scientific data in preparation for the Apollo landings. All five Lunar Orbiters were successful. They made more than 6,000 orbits of the Moon and photographed more than 99 percent of the lunar surface. The first three provided sufficient coverage for selection of eight candidate sites for Apollo, the fourth supplied detailed coverage of the front side of the Moon, and the fifth supplemented data provided by the others. At the end of their lifetimes, all were commanded to impact the Moon so as not to interfere with manned spacecraft.

Lunar Orbiter 1
August 10, 1966
First US spacecraft to orbit another planetary body. Returned medium and high resolution photos of nine primary and seven potential Apollo landing sites; crashed on Moon, October 29, 1966.

Lunar Orbiter 2
November 6, 1966
Returned 211 frames (422 medium and high resolution pictures); crashed on Moon, October 11, 1967.

Lunar Orbiter 3
February 5, 1967
Returned 211 frames including photographs of Surveyor 1; crashed on Moon, October 9, 1967.

Lunar Orbiter 4
May 4, 1967
Returned 163 frames; crashed on Moon, October 6, 1967.

Lunar Orbiter 5
August 1, 1967
Returned 212 frames including five Apollo sites, and provided near-lunar micrometeoroid data. Crashed on Moon, January 31, 1968.

Apollo

Apollo 11–Apollo 17
July 1969–December 1972
During six missions, astronauts landed and studied the lunar surface, collecting and bringing back lunar rock and soil samples and leaving instruments to study the Moon's interior.

Findings: (1) The Moon is a complex, evolved planet with three basic rock types and no life, past or present. (2) It is slightly egg-shaped, with the small end pointing toward Earth; consists of a crust, mantle, and perhaps a metallic core; is seismically inactive; and has unexplained fossil magnetism in the rocks, although the Moon has no magnetic field. (3) The rocks are igneous or derived from igneous rocks; ranging in age from 3 billion to 4.6 billion years, they are generally like those of Earth in chemistry and minerals, but are deficient in volatile elements such as hydrogen, sodium, and potassium; three new minerals never found on Earth were discovered (tranquillityite, armalcolite, and pyroxferroite). (4) Bombardment by cosmic dust seems to have occurred at a constant rate over the last few million years; rocks have remained exposed on the lunar surface for as long as 500 million years without being destroyed; the solar wind striking the Moon has a higher hydrogen-helium ratio than the Sun itself; no major changes in the intensity of solar flares and composition of particles erupted from them over the past 100,000 years; and the flux of galactic cosmic rays has apparently been constant.

Model of Surveyor.

Thin section of a lunar sample collected on the Apollo 14 mission.

The Planets

Planetary studies by unmanned spacecraft have been flybys, orbiters, or landers.

The first planet in the solar system, Mercury, was the focus of the Mariner 10 mission which gave scientists their first close look at the planet during three flybys. The spacecraft reached Mercury with a gravitational assist from Venus.

Venus was the objective of Mariners 1, 2, 5, and 10, though only 2, 5, and 10 were successful. Later, Pioneer Venus 1, an orbiter, and Pioneer Venus 2, a multiprobe, were a six-spacecraft effort to study the second planet from the Sun and closest one to Earth.

Mars was approached by both Mariner and Viking spacecraft. Viking landers reached the planet's surface in 1976, achieving the first soft landing of a spacecraft on another planet.

Pioneer and Voyager spacecraft were sent to Jupiter. Voyager 1 reached Jupiter in March 1979 and Saturn in November 1980. Its planetary mission completed, it continues to search for the edge of the solar system.

Voyager 2 encountered Jupiter in July 1979, Saturn in August 1980, and is now en route to a rendezvous with Uranus in January 1986 and Neptune in August 1989.

Thus, by the end of this decade, all planets in the solar system will have been visited except Pluto.

Mariner

Mariner 2
August 1962
First successful interplanetary probe, a flyby of Venus; data led to an accurate determination of the planet's mass, and measured high temperatures.

Mariner 3
November 1964
Mars flyby. Shroud failed to jettison and communications were lost with the spacecraft.

Mariner 4
November 28, 1964
Mars flyby. Reached the planet July 14, 1965. Photographed a heavily cratered, moonlike surface. Found that the Martian atmosphere is thin, with less than 1 percent the pressure of Earth's atmosphere, and is composed largely of carbon dioxide.

Mariner 5
June 1967
Venus flyby. Found weak magnetic field and very dense atmosphere.

Mariners 6 and 7
Launched February 25 and March 27, 1969. Mars flybys. Closest approach to planet achieved on July 31 and August 5, 1969.

Findings: Nitrogen is virtually absent from the atmosphere, solid carbon dioxide ("dry ice") occurs in the clouds and near the polar caps. and the dust particles in the Martian atmosphere probably consist of silicate materials derived from the planetary surface.

Mariner 8
May 1971
Mars orbiter mission. Launch vehicle failed.

Mariner 9
May 30, 1971
First Mars-orbiting spacecraft, November 1971. Transmitted 7,400 pictures of 100 percent of Martian surface as well as the planet's small moons, Phobos and Deimos.

Results: (1) Mars is a two-part world, with an ancient cratered surface in the Southern Hemisphere and a geologically younger surface, with volcanoes, canyons, and dry river channels, in the Northern Hemisphere. (2) It has a huge canyon, some enormous volcanoes, and sinuous channels which appear to be former river beds. (3) Landforms resembling lava flows occur in flat regions. (4) Layered deposits in the Martian polar regions suggest glacial periods in past times. (5) Solar ultraviolet light is not absorbed by the atmosphere and reaches the surface of Mars. (6) Periodic global dust storms were observed. (7) Phobos and Deimos are very dark and have irregular shapes and cratered surfaces.

Mariner 10
November 3, 1973
Venus (February 5, 1974)/Mercury (March 29, 1974) flyby.

The trajectory around the Sun swung it back for a second encounter with Mercury in September 1974 and for a third in March 1975.

Data from Venus: (1) No significant magnetic field; (2) a notable disturbance in the solar wind is produced as it flows past Venus; (3) it is closer to a perfect sphere than Earth is; (4) ultraviolet images of the atmosphere revealed streamline and circulation patterns, including Y- and C-shaped structures; (5) the upper atmosphere rotates much more rapidly (once in about 4 Earth days) than Venus itself; (6) hydrogen and helium were detected in the atmosphere.

Data from Mercury: (1) Photographs showed desolate landscapes, remarkably similar to the Moon's with huge craters; long narrow valleys; a feature unique to Mercury, long scarps, or cliffs; flat plains; and a huge circular impact basin (Mare Caloris) about 13,000 km (810 mi) in diameter; (2) it's closer to a perfect sphere than Earth is; (3) not only the smallest planet, but the densest with a metal-rich core; (4) a tenuous atmosphere includes exotic gases such as argon, neon, and helium; (5) its magnetic field is about a tenth as strong as Earth's; and (6) the temperatures are extreme, from 425°C (770°F) to −183°C (−297.4°F).

Pioneer

Pioneer 10 (Pioneer Jupiter)
March 3, 1972
Jupiter encounter, December 3, 1973

First spacecraft to leave the solar system, 5:00 a.m. PDT, June 13, 1983. Carries plaque with an easily-interpreted message: drawing of a man and a woman, a diagram of the solar system, and a map locating the solar system with reference to some galactic pulsars.

Basic mission was the first flyby of Jupiter. In addition, was the first flight beyond Mars and first crossing of the asteroid belt; first close-up pictures of Jupiter's Great Red Spot and atmosphere; and first crossings of orbits of Uranus, Pluto, and Neptune.

Discoveries: (1) The heliosphere (Sun's atmosphere) extends much farther than previously thought and appears to "breathe" in and out once every 11-year cycle; (2) Jupiter is a liquid planet; (3) first model of Jupiter's huge magnetosphere; (4) first accurate measurements of mass and densities of Jupiter's planet-sized moons; (5) proof of origin of the gegenschein and zodiacal light.

Pioneer 11 (Pioneer Saturn)
April 6, 1973
Jupiter encounter December 2, 1974

Carries plaque identical to Pioneer 10. Passed within 42,760 km (26,725 mi) of the planet's cloud tops taking the only existing pictures of its polar regions. Jupiter's gravitational field was used to swing it back across the solar system to Saturn.

Saturn encounter, September 1, 1979

Renamed Pioneer Saturn after Jupiter encounter. 565 new discoveries came from its path through the ring plane [2,000 km (1,200 mi) below them] and within 21,400 km (13,300 mi) of the cloud tops.

Results: The planet has a magnetic field, magnetosphere, and radiation belts; its core is about twice the size of Earth; its magnetic field is 1,000 times stronger than Earth's; it appears to have more and narrower belts and zones than Jupiter; identified two new rings and found an 11th moon; measurements of Titan discouraged evidence for possibility of life. The data was useful for planning encounters of Voyagers 1 and 2.

Pioneer Venus 1
May 20, 1978
Reached Venus, December 4

Orbiter mission to observe Venus for one complete rotation on its axis; globally surveyed its atmosphere and environment, studied its topography, calculated its shape and density.

Results: (1) First full-disc picture of Venus shows a turbulent, cloudy atmosphere, bright cloud areas wrapped about both polar regions, and a Y feature covering most of the central

Artist's concept of Pioneer 10 leaving the solar system.

Artist's concept of Aphrodite, Venus' largest highland region, with an outline of the continental United States.

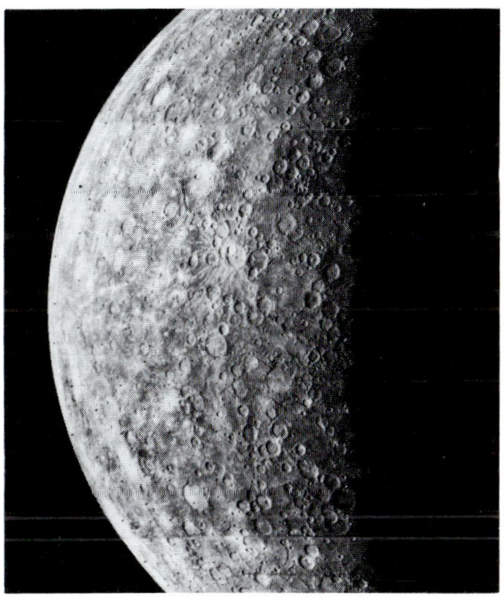

Photomosaic of 18 pictures of Mercury taken by Mariner 10 when the spacecraft was 200,000 kms (124,000 mi) distant on its approach to the planet March 29, 1974.

part of this disc; (2) a thick, pale yellow opaque atmosphere obscures the surface, but radar scanner revealed flat rolling plains, a mountain as high as Mount Everest, great rift valleys, continent-sized highland areas, and two large volcanoes; (3) data provided measurements on high-speed winds, changing global patterns of clouds and cloud-level winds, a high-altitude haze of sulfuric acid, and a surface temperature of 482°C (900°F).

The Orbiter's mission has been extended to 1985.

Pioneer Venus 2
August 8, 1978
Reached Venus, December 9

A multiprobe mission made up of a Bus, a Large Probe, and three identical Small Probes to measure the atmosphere top to bottom. The Bus measured the upper atmosphere and then burned up. The Probes descended to the surface; not designed to survive impact, but one returned data for approximately 67 minutes.

Data showing the presence of large amounts of rare gases in the atmosphere suggest a far larger contribution by the Sun to Venus' atmosphere than to Earth's during the early evolution of the solar system.

Viking

Viking was designed to orbit Mars and land and operate on its surface. Two identical spacecraft consisting of an orbiter carrying a sterilized lander were launched in 1975 from Cape Canaveral and cruised through space for almost a year.

Viking 1
August 20, 1975
Began orbiting Mars, June 19, 1976
Landed July 20

Viking 2
September 9, 1975
Began orbiting Mars, August 7, 1976
Landed September 3

Viking's primary mission ended November 15, 1976, 11 days before Mars passed behind the Sun. After conjunction in mid-December 1976, telemetry and command communications were reestablished, and the extended mission operations began. Orbiter 1 continued working until the summer of 1980. Orbiter 2

Full-disc picture of Venus taken from 65,000 kms (40,000 mi) by the Pioneer Venus Orbiter on February 10, 1979.

Artist's concept of the Viking launcher mission profile showing, left to right: bloshield separation, lander capsule separation, the descent parachute deployment, terminal propulsion, and entry to landing.

View of Jupiter showing never-previously-seen aspects of the planet's cloud tops taken by the Pioneer 10 spacecraft as it flew past the giant planet in December 1973, from 1,840,000 km (1,121,000 mi) away (above left).

Jupiter's Great Red Spot and surrounding region photographed by Voyager 1 on March 1, 1979 from a distance of 5 million km (3 million mi) (below left).

First color picture taken on the surface of Mars by the Viking 1 Lander (above). The scene, covering about 67′ from left to right, was scanned three times, each time with a different color filter. The color was reconstructed with computer processing.

(Right), global color images of the four Galilean satellites taken by Voyager 1: Io (upper left), Europa (upper right), Ganymede (lower left), and Callisto (lower right).

ended its mission in July 1978, and Lander 2 in April 1980. Lander 1 was the longest-lived; it operated until November 1982.

With but one exception, the science instruments acquired more data than was expected: (1) Biology experiments analyzed Martian soil and discovered chemical activity, but no evidence of life; (2) measurements were made of some physical and magnetic properties of the soil and the composition and physical properties of the upper atmosphere; (3) there was nearly continuous monitoring of weather at the landing sites where surface temperatures ranged from 29°C (−20°F) in the afternoon to −84°C (−120°F) at night; (4) there were dust storms, but wind velocities were low; (5) the north polar ice cap is composed of water ice.

The first color photographs showed the Red Planet truly red, the color due to oxidized iron. The total number of pictures exceeded 4,500 from the landers and 52,000 from the orbiters. The landers provided the first close-up look at the surface, monitored variations of atmospheric opacity over a full Martian year, and determined the mean size of the aerosols. The orbiter cameras observed new terrain and provided clearer detail on known features, including some color and stereo observations, and mapped about 97 percent of the surface.

Voyager

The Voyager missions to Jupiter and Saturn were built on the data previously acquired during the Pioneer 10 and 11 flybys of Jupiter. The photographs of the planets—their atmosphere, rings, and satellites—surpassed anything scientists expected. With the scientific data they accumulated, the Voyagers transformed completely our view of and knowledge about the giant planets.

Each Voyager carries a "Sounds of Earth" record made of copper and gold-plated. The records contain two hours of sounds and music, digital data containing pictures, and a digital message from President Jimmy Carter.

Voyager 2
August 20, 1977

Voyager 1
September 5, 1977

Jupiter encounter, March 5, 1979

Launched after Voyager 2 but placed on a faster trajectory, Voyager 1 passed it and reached Jupiter four months earlier.

Flew by within 280,000 km (172,750 mi) of the atmosphere. Observations: (1) The colorful striped, constantly moving atmosphere has dozens of storms and complex cloud structures; (2) discovered a torus (doughnut-shaped tube) circling the planet, a plasma cloud of intense heat that encloses the orbit of Io; (3) discovered a thin, flat ring of dark particles; (4) aurorae fill Jupiter's skies and there are intense, bright lightning strikes; (5) the Great Red Spot is surrounded by rolling clouds and spiral streams of clouds; (6) the Galilean satellites were the first found to have color: Io, bright orangish, has active volcanoes; Europa, amber, has long linear streaks crossing its surface; Ganymede, brown, larger than Mercury, has impact craters with right ray systems; and Callisto has a basin surrounded by a series of concentric rings.

Saturn encounter, November 12, 1980

Only 19 km (12 mi) off course, observations at Saturn were equally surprising: (1) The atmosphere is similar to Jupiter's with dark and light cloud markings and swirls, eddies, and curling ribbons; but the belts and zones are more numerous and a thick haze mutes the

markings; (2) the planet's rotation period is 10 hours. 39 minutes, 26 seconds; (3) the rings became likened to the grooves of a phonograph, rings within rings numbering in the hundreds; there were spokes in the B ring and shepherding satellites controlling the F ring; (4) Titan, the satellite almost as large as Jupiter's Ganymede, has a dense haze that hides its surface; (5) the inner five satellites—Mimas, Enceladus, Tethys, Dione, and Rhea—are mainly water ice and each has distinctive features; of the outer satellites, Iapetus has one bright, one dark hemisphere; (6) the satellite count went up to 15.

Voyager 2

Jupiter encounter, July 9, 1979

Concentrating on areas and characteristics of the planet and its satellites that complemented the findings of Voyager 1, Voyager 2 brought new images and scientific data: (1) Confirmed the constant changes of the atmosphere and changes in the Red Spot in size and color; (2) photographed the ring above and below to find it 6,500 km (4,000 mi) wide and 10 km (6 mi) thick and a fainter ring within the inner edge of the bright ring; (3) a time-lapse movie was made of 150 frames to record the volcanic activity on Io, another view of Callisto showed uniform impact crater distribution, and Europa's flat surface came to resemble a cracked egg shell.

Saturn encounter, August 25, 1981

Just 2.7 seconds early and only 48.27 kilometers (30 miles) from the aim point, Voyager 2 approached Saturn from above the rings. Its instruments had been adjusted to take advantage of knowledge gained from Voyager 1 and, 23,000 km (14,294 mi) closer, concentrated on selected targets.

A closer study of atmospheric motion showed new features: cloud vortices (small hurricanes), high-speed jet streams, eddies at higher latitudes. Its temperature ranges from 80°K to 95°K (176°F–203°F) at the cloudtops.

It discovered a plasma torus around the planet with temperatures 300 times hotter than the Sun's corona and twice as hot as the torus around Jupiter.

Better cameras revised the ring count to thousands and a series of time-lapse movies studied the B ring spokes. Close looks at several satellites showed distinctive features and high-resolution photographs were taken of seven of the newly-discovered satellites.

Kuiper Airborne Observatory

March 1977

The Kuiper Airborne Observatory, a modified C-141 aircraft, carries the world's largest airborne telescope. Scientists discovered the rings of Uranus while observing the temporary disappearance of a faint star behind the planet. Five rings lie 18,000 km (11,000 mi) from the planet's cloud tops; they appear to consist of four thin inner rings that follow nearly circular orbits around the planet, and one thick outer ring whose orbit may not be exactly circular.

Sounding Rockets

The first adventures into space were with suborbital sounding rockets and the Sounding Rocket Program continues to conduct scientific experiments that require only a few minutes duration above the atmosphere. Their effective lifetime—when all scientific data is collected—is the few minutes before dropping back to Earth.

They travel in nearly vertical trajectories, may have one or more stages, carry packages of instruments to between 48 and 161 km (30–100 mi), and return data by telemetry or capsule. They are the only vehicles effective at those heights because balloons cannot carry equipment above 48.27 km (30 mi) and satellites are impractical below 160.9 km (100 mi). A few sounding rockets reach altitudes of 6,436 km (4,000 mi). Those designed for lower altitudes investigate properties of the upper atmosphere; at higher altitudes they have contributed to high energy astrophysics and solar astronomy.

Sounding rockets are used to test prototype instruments for satellites, to provide scientists with instruments at the precise time and place needed, and to support observations by satellites.

Possible variations in the chemical composition of Saturn's ring system are revealed by this computer-enhanced color composite of two images transmitted by Voyager 2 on August 17, 1981, from 8.9 million km (5.5 million mi).

For the Classroom

1. Research topics:
   - Planetary images taken by spacecraft vs. those taken by Earth-based telescopes
   - Compare the old and new information about a planet that has been visited by a spacecraft
   - Infrared and ultraviolet astronomy
   - Robotics
2. Collect images of the planets and their satellites taken by spacecraft and compare them with pictures in pre-Space Age books.
3. Encourage students to construct and test robotic devices for science fair projects.
4. Have your students investigate local use of robots. If they are being used, arrange a field trip.
5. Design a robot for future planetary missions—to land, collect surface samples, and return the samples to Earth.
6. Teachers, Grades 6–12, interested in becoming certified for use of the Lunar Sample Educational Packet should contact their Center Educational Programs Officer. (See Appendix.)

IX

Technology Utilization

Technology—technical "know-how"—is almost limitless in its adaptability and uses.

During 25 years of aerospace research, NASA's programs have required and produced technological advances. The resultant technologies are diverse and have accumulated to become an important national asset.

One of the demands of the agency's mandate is the widest possible use of this resource. Because technology is transferable and there is always a potential for new applications, NASA's Technology Utilization Program was established to provide the link between the technology and the user.

Earlier chapters outlined the NASA programs that have generated much of today's new technology. This chapter introduces a selection of products and processes, called spinoffs, that came from secondary application of that technology.

A technician inspects the electronic circuitry of a QuadraScan® Longterm Flow Monitoring System provides information on the water flow of city sewer systems.

Computers

Computer processing is having a profound impact on American business and industry. Computerization of the private sector is expected to continue growing throughout the 1980s.

With its wealth of computer resources developed in the aerospace program, NASA is uniquely equipped to aid businesses as they make the transition to the computer age.

Computer programs, or software, are sets of instructions that tell a computer how to retrieve its stored information for the desired results. Developing new programs is expensive and time-consuming. Often, however, a program developed for one application can easily be adapted to a new purpose. Thus many of NASA's computer programs can be used to great advantage in business and industry.

Through the Computer Software Management and Information Center (COSMIC)® at the University of Georgia, NASA makes software from space programs available to the private sector at a fraction of the original cost. Thousands of programs have been distributed, resulting in the savings of millions of dollars.

In one such instance, Shell Oil Company of Houston, Texas, adapted a COSMIC program for use in the production of chemicals for plastic products. Computer analysis of chemical compounds helps scientists predict how new composite structures will perform when applied to such products as automobiles, appliances, and children's toys. The COSMIC program was used to evaluate the accuracy of the company's new computer code.

Rohr Industries, Inc., Chula Vista, California, specialists in manufacture of nacelles, has made extensive use of a COSMIC program, saving six man-months of programmer time necessary to develop alternative software. Since 1979, the Information Systems Department of Illinois Bell Telephone has been using a COSMIC inventory providing aids and refinements for structured assembler language programming techniques.

COSMIC has more than 1,500 computer programs available, which are serving a variety of fields from developing huge gears for wind turbines to bank record processing to analyzing blood samples in medical research laboratories.

A glass-enclosed subway station in Toronto, Canada, was made possible by a unique glazing concept developed by PPG Industries.

The NASA Structural Analysis (NASTRAN)® computer program has also found wide application in industry. This general purpose program analyzes a design and predicts how it will stand up to vibration and other forms of stress.

At its glass research center, PPG Industries of Pittsburgh, uses the NASTRAN computer program to analyze the stability of enclosures made entirely of glass, to simulate stresses on large containers of molten glass, and to analyze stress effects of solar heating on flat glass.

Ingalls Shipbuilding Division of Litton Industries, Pascagoula, Mississippi, a leading designer and producer of Navy combat ships, oil drilling rigs, barges, and other vessels for the offshore marine industry, has used the NASTRAN program in shipbuilding.

Fabric Structures

In 1967, NASA was searching for a fabric for astronaut space suits. Owens-Corning Fiberglas Corporation of Toledo, Ohio, produced a glass fiber yarn that was woven into a fabric and coated with Teflon®. This fabric was subsequently developed into heavier versions and used to construct permanent tent-like roofs on large public buildings. The Silverdome, home of the National Football League's Detroit Lions in Pontiac, Michigan, is one of the best-known of these which include department stores, university recreation centers, and Sea World's Florida Festival in Orlando.

The Silverdome, home of the NFL's Detroit Lions in Pontiac, Michigan, is an air-supported Teflon-coated Fiberglas fabric roof for year-round utility.

Consumer Products

Space technology has led to a variety of consumer products. A NASA engineer, who helped design cooling systems for space suits, with some of his colleagues has developed a line of liquid-cooled sportswear for joggers and other athletes. Known as Techni-Clothes, they use heat-absorbing gel packs and are made to allow more strenuous activity without overheating.

Another entrepreneur used NASA technical information from the Space Shuttle program to develop his own water filter. About the size of a thermos bottle, the filter attaches to a faucet and removes chemical tastes and odors from tap water.

A third businessman repairs plastic canoes used for whitewater canoeing with an Inductron Toroid Welder, an advanced welder that was developed at the Langley Research Center for use in space to repair plastics and other synthetic materials.

These New York marathoners are doing warmups, wearing the Techni-Clothes cooling headbands.

Health and Medicine

In the space program, controlling fluids has been Important in the huge liquid-propellant rockets as well as in the tiny automated equipment that scooped up Martian soil. This sophisticated fluid control technology is now being applied in medicine: an artificial sphincter for urinary control, an implantable system to deliver internal medication automatically, and a blood filtration system for the treatment of rheumatoid arthritis.

A portable Medical Status and Treatment System developed for use in remote areas incorporates astronaut-monitoring, electronic circuitry, and microminiaturization. The Ames Research Center assisted in the development of a crawling aid for brain-injured children.

NASA's computer systems are being used in medical research. The Centers for Disease Control (CDC) in Atlanta, Georgia, use a COSMIC program to analyze data from tests of human body substances. A medical laboratory in Maine uses NASA computerized technology to develop new diagnostic techniques.

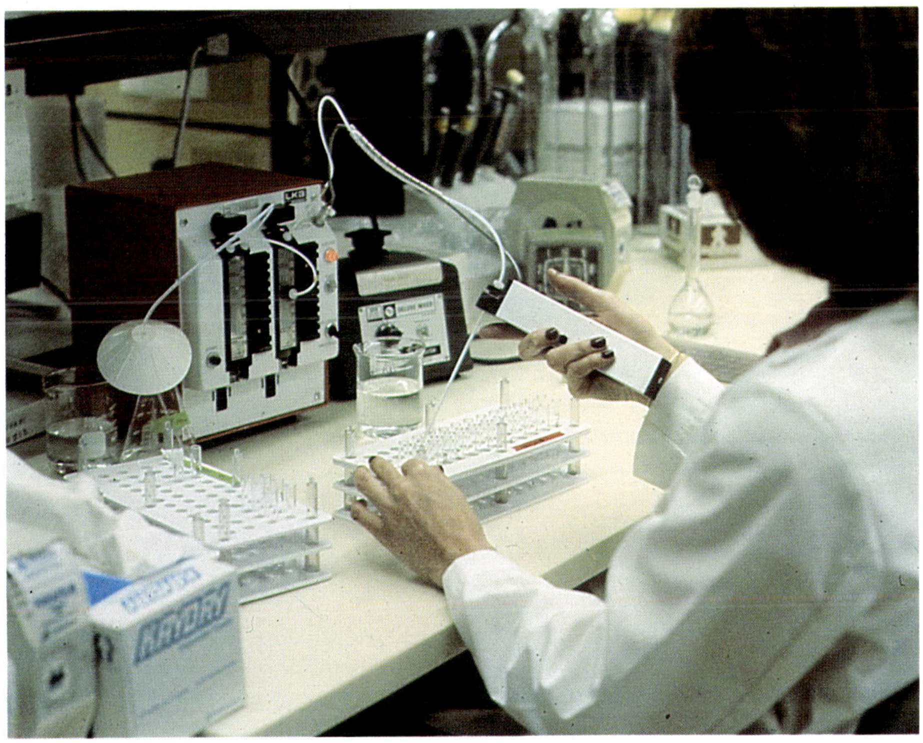

A chemist at the Centers for Disease Control in Atlanta where the COSMIC exchange is used for many medical research programs.

Industry

At a time when worker productivity is dramatically affecting the economic futures of many industries, NASA technologies are being brought in to help. The Cool Vest™ is an example of a spin-off product that can increase worker productivity. This lightweight cooling garment can be worn by workers in high temperature environments to allow longer working periods. Another NASA-supported device is the automatic welding system, designed to replace manual welding with a technology higher in productivity and lower in cost.

One superefficient insulation developed to keep the Saturn V's fuel tanks cold was a spray-on polyurethane foam technique devised by Rockwell International Corp.; the technique found commercial application as insulation for the storage wells on tuna boats.

NASA technology developed for accurate pressure measurements in wind tunnels is being applied in industrial plants to automatically control individual valves and actuators. The DPT6400, produced by Pressure Systems Incorporated, is an outgrowth of electronically scanned pressure (ESP) technology developed at Langley Research Center.

For several years some building construction has been based on a money-saving method of preparing building specifications derived from NASA technology; the technique was developed to obtain quality construction while holding down cost of launch facilities, test centers, and other structures.

Environment

Environmental and public safety management are more effective as a result of space age technologies. Sewer monitoring has long been an inexact practice, resulting in either inadequate or redundant sewage treatment facilities. A new monitoring system called QuadraScan borrows satellite sensing technology to take accurate readings of water flow in a sewage system.

Fire safety has also been enhanced by NASA spinoffs. Durette®, a fire-resistant fabric developed for the oxygen-enriched (hyperbaric) atmospheres of spacecraft, has a variety of Earth applications: researchers on deep diving and oxygen therapy use Durette in their hyperbaric chambers, auto racers wear Durette suits, and Durette filters clean emissions from a high-temperature furnace at a metal products company.

High noise levels can pose problems in any industry, but the poultry industry has a special problem: materials used to absorb noise must be washable and must not allow bacterial growth. In a three-year program, Georgia Tech's Engineering Experiment Station (EES), supported by NASA and the Georgia Department of Agriculture, developed a sound-absorbing panel for use in poultry plants. EES uses a three-inch fiberglass core covered by a polyester film similar to that used in space for vapor protection. The results of the EES demonstration are considered applicable to other food processing industries where similar sanitary constraints exist.

Tuna boats use a NASA-developed spray-foam insulation for improved refrigeration.

For the Classroom

1. Have your students compile a list of consumer products from the space program that they find in daily use.
2. Ask your students to find out if any of the community benefits (environmental, health, industrial) listed in this chapter are found in their area; how are they used? if not, could they be ?
3. Have your students research new products, other than those mentioned in the text, that had a beginning in NASA programs. Do they have local, national, or international applications?
4. Divide the class into special interest groups—sports, architecture, medicine, environment—to prepare in-depth reports on benefits from NASA research.

X

Tracking and Data Systems

Pioneer 10 at the edge of the solar system. Columbia and Challenger crews in space. Landsat. Voyager encounters with Jupiter and Saturn. Lunar bases. The vital link between these missions and Earth is the Space Tracking and Data Systems.

Their name describes their work—to keep track of where spacecraft are in orbit, tell them what to do, get information from them, and process that information into a meaningful form. This is done primarily through two worldwide networks—one for Earth-orbiting operations and the other for deep space missions. A global communications system called NASCOM (NASA Communications System) links the networks with NASA mission control centers.

The networks support an average of 20 to 30 different satellites and space probes daily. Through the years they have enjoyed some remark-

The 64-meter Deep Space Network antenna at the Madrid Tracking Station Complex.

Workmen standing in the reflector of the 64-meter diameter antenna of the Jet Propulsion Laboratory's Deep Space Network station at Goldstone, California.

able successes as they have adapted to changing requirements and technological progress.

Originally able only to receive and transmit data, tracking stations now also have advanced data processing and command capabilities. Today there are two separate tracking systems. The Space Tracking and Data Network (STDN), managed by the Goddard Space Flight Center (GSFC), is an international system of tracking facilities for Earth-orbital and suborbital missions. The Deep Space Network, managed by the Jet Propulsion Laboratory (JPL), consists of three tracking stations around the world for the support of deep space missions.

The new Tracking and Data Relay Satellite System (TDRSS), with the first satellite launched from the Space Shuttle Challenger on STS-6 in April 1983, is an in-orbit communications link with other spacecraft and eventually will replace ground-based STDN.

Minitrack 1958–1963

In 1958, NASA acquired the inter-

Blossom Point, Maryland, Tracking Station showing (from left): an experimental 16-element Yagi antenna; Minitrack antennas (next to building); and the Operations Building.

national network of "Minitrack" stations that had been formed by the Naval Research Laboratory to support the International Geophysical Year. These stations could receive, record, and transmit telemetry data, but could not give commands. Between 1959 and 1963, NASA added new stations to support the growing space program and upgraded the capabilities at existing stations. In 1962, construction of the first 26-meter (85-foot) diameter parabolic antenna was completed at Fairbanks, Alaska.

Minitrack Stations:
1958
San Diego, CA
Blossom Point, MD
Antiqua, West Indies
Quito, Ecuador
Lima, Peru
Antofagasta, Chile—to 1963
Santiago, Chile
Woomera, Australia
Esselen Park, South Africa
Johannesburg, South Africa
1959
Fort Myers, FL
1960
East Grand Forks, MN
Goldstone, CA
1961
Fairbanks, AK
St. John's, Newfoundland
Winkfield, England
1963
Rosman, NC
Canberra, Australia

Satellite Network 1963–1964

As 26-m antennas were added to the Minitrack system in the early 1960s, the system became known as the Satellite Network. In 1964, 12-m (40-ft) antennas were added in Johannesburg, South Africa; Quito, Ecuador; and Santiago, Chile.

Space Tracking and Data Acquisition Network (STADAN) 1964–1972

In 1964, NASA installed the Satellite Telemetry Automatic Reduction (STAR) system, a data processing system that significantly expanded the satellite network's capabilities. The network, which became known as STADAN, could command satellite functions and acquire data, as well as track satellites. Increased capabilities of the STADAN system allowed some Minitrack stations to be phased out. At the end of 1969, ten STADAN stations were operational.

Project Apollo tracking ship, *USNS Mercury,* at Port Hueneme, California.

STADAN Stations:
Canberra (Orroral Valley), Australia
Fairbanks, AK
Fort Myers, FL
Goldstone, CA
Johannesburg (Hartebeesthoek), South Africa
Quito, Ecuador
Rosman, NC
Santiago, Chile
Tananarive, Madagascar
Winkfield, England

Manned Space Flight Network (MSFN) 1962–1972

There was a separate tracking network for manned spaceflight until 1972. The network created in 1958 to support Project Mercury was augmented in 1962 for Project Gemini. The MSFN stations consisting of airplane, ship, and ground-based antennas could track, command, receive data, and communicate with the astronauts and the target vehicle. The system was updated again for the Apollo program and consisted of ten 9-meter antennas, one 9-meter transportable station, five ships, and eight aircraft.

Nigerians watch twin-helix antennas at Kano, an MSFN station, swing to acquire a Mercury capsule.

MSFN Stations:
Antigua
Ascension Island
Bermuda
Canary Islands
Canberra, Australia
(Honeysuckle Creek)
(Orroral Valley)
Canton Island
Carnarvon, Australia
Corpus Christi, TX
Eglin AFB, FL
Goldstone, CA
Grand Bahama Island
Grand Turk Island
Guam
Guaymas, Mexico
Kano, Nigeria
Kauai, Hawaii
Madrid, Spain
Merritt Island, FL
Muchea, Australia
Point Arguello, CA
Tananarive, Madagascar
White Sands, NM

In 1972, MSFN was merged with STADAN.

Spaceflight Tracking and Data Network (STDN) 1972–to date

STADAN and MSFN became STDN, an international network of 15 stations. Twelve of these stations track manned and unmanned Earth-orbital and suborbital missions. Three special purpose stations will be used for support of the Space Shuttle. During the 1970s, the STDN network was continually improved to provide greater data processing capabilities.

The STDN system provides coverage up to about 20 percent of the time. Most equipment on the STDN network was installed in the mid-1960s to support the Apollo program, and although obsolescence and maintenance difficulties have increased with time, the network has continued to provide consistent service longer than expected. The new TDRS system will allow many STDN stations to be closed.

The Tracking and Data Relay Satellite System (TDRSS) Space Ground Link Ku-Band Antenna System, at White Sands, New Mexico.

STDN Stations:
Ascension Island
Bermuda
Canberra (Orroral Valley), Australia
Fairbanks, AK
Greenbelt, MD
Goldstone, CA
Guam
Kausi, HI
Madrid, Spain
Merritt Island, FL
Quito, Ecuador
Rosman, NC
Santiago, Chile
Winkfield, England

Deep Space Network (DSN)

The Deep Space Network is a system designed to provide command, control, tracking, and data acquisition for deep space missions. Its three sites in Goldstone, California; Madrid, Spain; and Canberra,

Australia are located approximately 120° apart and provide 24-hour line of sight coverage.

DSN, managed by JPL, consists of three 64-m (210-ft) diameter antennas, five 26-m (85-ft) antennas, and one 34-m (111-ft) antenna. During the Voyager 1 encounter with Saturn, DSN recovered over 99 percent of the 17,000 images transmitted. The network was able to determine the position of Voyager 1 to within 337 km (209 mi) upon its closest approach to Saturn. This high level of performance was made possible with the use of the network's radiometric system, the spacecraft cameras, and the use of antenna arraying. The arraying technique is done electronically by combining signals received from two antennas at each site.

Tracking and Data Relay Satellite System (TDRSS)

When operational, the in-orbit tracking system, TDRSS, will revolutionize global coverage of Earth-orbiting spacecraft. The largest and most advanced communications satellites developed thus far, TDRSS will consist of two satellites and an in-orbit spare. They will provide almost full-time coverage of the Space Shuttle and up to 26 other satellites.

The TDRSS satellites weigh about 2,250 kilograms (5,000 lbs) and measure 17 meters (57 ft) across.

TDRSS CONCEPT

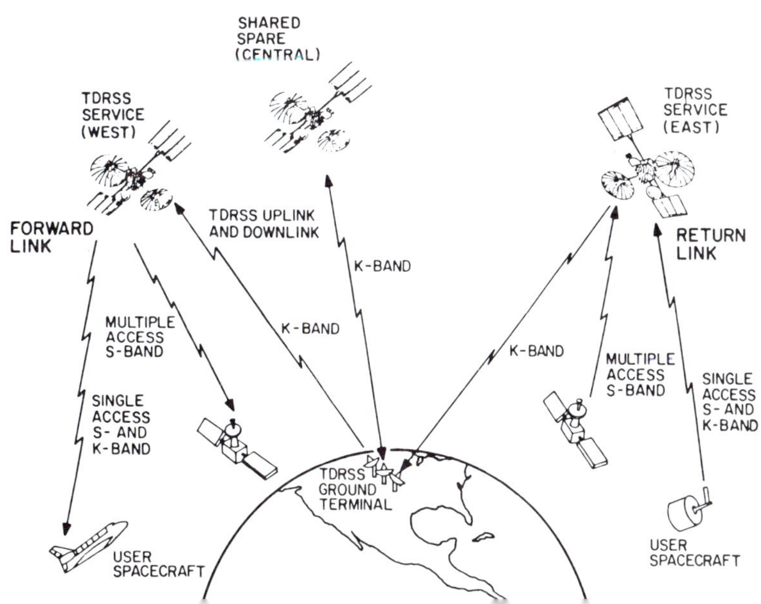

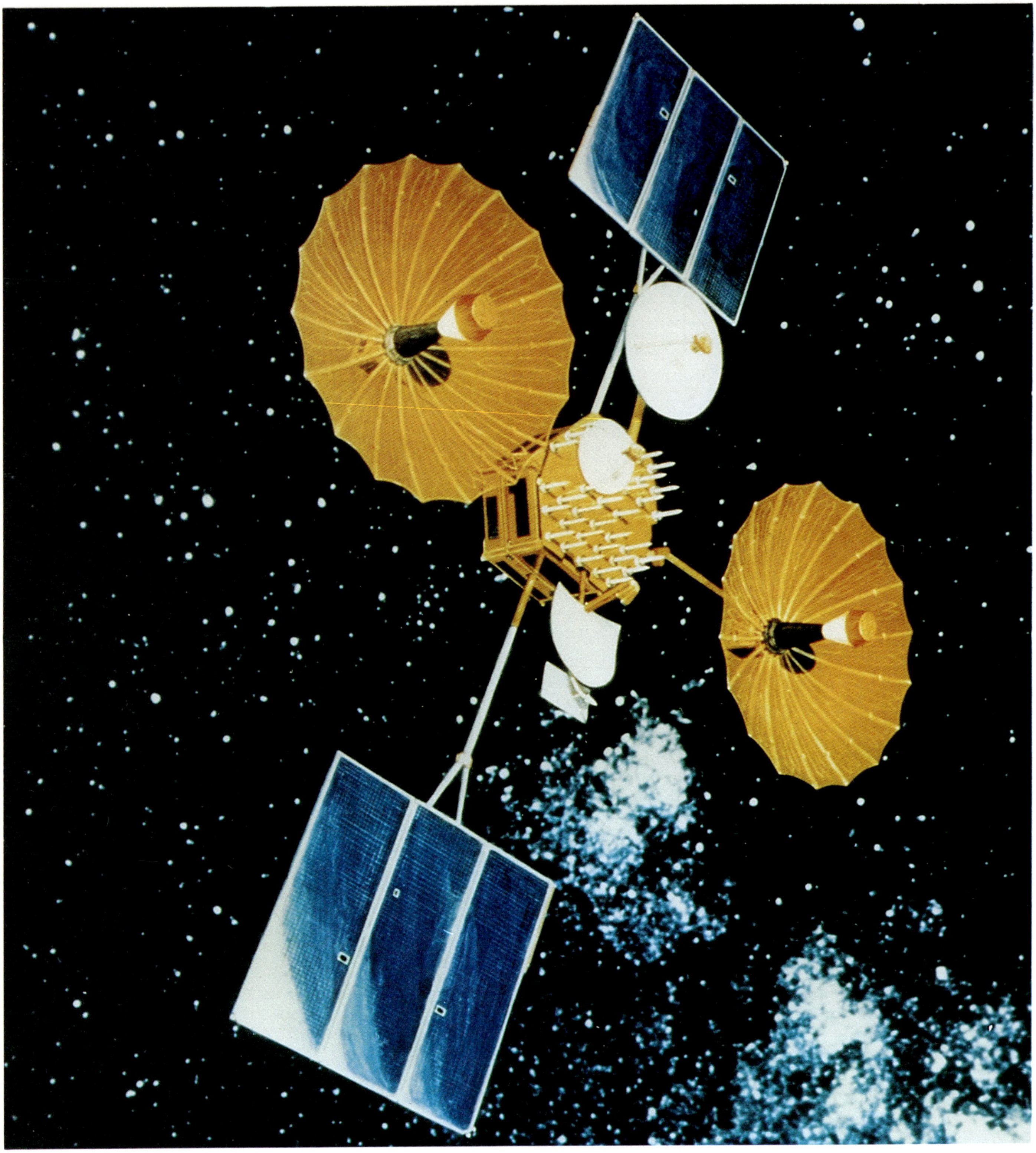

The operational satellites will be positioned over the Equator about 130 degrees apart, with the spare centrally located for use in case of a malfunction.

The data acquired will be sent directly to NASA's White Sands Test Facility in New Mexico. Facilities at the White Sands ground terminal include three 18-m (59-ft) Ku-band communications.

Initially, TDRSS will support the Space Shuttle, Spacelab, and Landsat 4.

Artist's concept of TDRS. The spacecraft bus is in the center; from it project the two steerable dish antennas, which are five meters (16 ft) in diameter; the solar panels on booms extending perpendicularly to the big dish axis; and the smaller white dish, which is the space-to-ground link.

For the Classroom

1. Have your students locate tracking stations on a map. Discuss why several stations are needed for one system, but only three for another.
2. Keep a class file on the TDRS System.
3. Have your students research one tracking station—its use, geography, history, impact on the community.
4. Secondary school teachers may obtain a copy of *Teachers' Guide for Building and Operating Weather Satellite Ground Stations* from the Educational Programs Officer, NASA Goddard Space Flight Center (202.3), Greenbelt, MD 20771. The publication gives the information needed to construct, modify, and operate a weather satellite recording station.

Epilogue

Perspectives, Plans, Prospects

On October 1 this year NASA will be 25 years old, a short time in the life of a nation but a quarter century of extraordinary advancement. Achievements of these 25 years are part of history. The exhilarating moments of accomplishments have become logical steps in progress.

Consider the legacy of this period:

- Apollo 11, the climax of humanity's long striving to break away from the confines of Earth to explore another world, and the beginning of a future that already has seen exciting advancements in knowledge and new discoveries in exploration. Ten more men landed on the Moon; a decade later astronauts and scientists are working in low Earth orbit.
- Pioneer, Mariner, Voyager, and Viking with cameras that extended our eyes to the outer planets, experiments that added immeasurably to our knowledge of the solar system, pictures that brought a new sense of wonderment, and discoveries that

Rockwell International concept of a space station. The picture also shows an Orbital Transfer Vehicle delivering a communications satellite to geosynchronous orbit from the space station.

brought once-distant worlds closer. Mars, Jupiter, Saturn, and their satellites are now places, familiar and unforgettable.

- Scientific satellites that have introduced us to the elements of Earth's immediate neighborhood, the Sun, and the baffling mysteries at the edge of the universe.
- Aeronautical research that leads the world.
- Applications satellites, now a part of our daily routine.
- Applications of space technology that have enriched our lives.

What Next—the 1980s

". . . the dream of yesterday is the hope of today, and the reality of tomorrow."
Robert H. Goddard
High school graduation talk, 1904

Aeronautics

An evolutionary science, the aeronautical progress made since 1958 is but the groundwork for the next generation of vehicles, the current research in laboratories, wind tunnels, and computers. 1984 will see tests of the X29A. Flight testing of the Laminar Flow Control program will continue through September 1986. The conceptual designs of the 1980s will become the aircraft of the 1990s.

1980s conceptual design of a 1990s commuter aircraft.

International

European Space Agency (ESA)

Spacelab 2
Scheduled for Shuttle launch, November 1984
Mission duration: 7 days
Objective: To demonstrate Spacelab's capabilities through a multidisciplinary research program and to verify system performance. First mission with a multipallet configuration, an igloo (a pressurized container attached to the first pallet and containing the avionics needed to operate and control all Spacelab subsystems and cargo bay experiments), and the ESA-developed Instrument Pointing System. (C)

Spacelab 3
Scheduled for Shuttle launch, September 1984
Mission duration: 7 days
Objective: First operational mission. Multidisciplinary, emphasizing investigations requiring the low-gravity environment of Earth orbit. Performance of equipment and facilities designed for life science investigations on Spacelab 4 will be evaluated. (C)

Spacelab 4
Scheduled for Shuttle launch, December 1985
(See Germany, below.)

International Solar Polar Mission (ISPM)
Scheduled for Shuttle launch, 1986
Designed to explore the polar regions of the Sun, ISPM will travel outside the ecliptic plane of the solar system. NASA will launch the satellite which will have a U.S. experiment on board, and provide tracking through the Deep Space Network. (C)

Space Plasma Laboratory
Scheduled for Shuttle launch, August 1988
Mission duration: 7 to 9 days
Objective: To conduct experiments in low Earth orbit using radio frequency transmitters and beam accelerators to perturb the near-space plasma environment; direct practical benefit will be in predicting the effects of geomagnetic storms on communications, power line grids, over-the-horizon radars. (C)

Germany

Active Magnetospheric Particle Tracer Explorers (AMPTE)
Scheduled for launch in 1984
AMPTE will study the interaction between the solar wind and Earth's magnetosphere through release of chemical tracer elements. For this dual spacecraft mission to be launched on the same Delta vehicle, NASA will provide the Charge Composition Explorer (CCE) and West Germany, the Ion Release Module. (C)

Space Biomedical Laboratory (Spacelab 4)
Scheduled for Shuttle launch, December 1985
Mission duration: 7 days
Objective: The mission is devoted to life sciences research related to the safety, well-being, and productivity of humans in space and to fundamental problems in gravitational biology. The common test subjects are the six human crew members, four squirrel monkeys, and 48 rats. (C)

Galileo
Scheduled for Shuttle launch, 1986
A two-spacecraft mission to Jupiter. Germany is contributing a Retro Propulsion Module to inject one spacecraft into Jovian orbit as well as other mission hardware. (C) (See Space Sciences, below.)

ROSAT
Scheduled for Shuttle launch, 1987
A NASA/German Federal Ministry for Research and Technology (BMFT) astrophysics mission; an X-ray telescope, it will produce an X-ray survey of the sky with detailed observations of specific sources. (C)

Space Science

Pioneers 10 and 11
After crossing the orbit of Neptune in June 1983, Pioneer 10 will continue to gather and relay detailed scientific information from previously unexplored outer reaches of the solar system. It is seeking the heliopause, the boundary where the solar wind hits interstellar gas.

Pioneer 11, also known as Pioneer Saturn, after successfully concluding its mission to the ringed planet, also headed out of the solar system to interplanetary space.

Voyager 1
Voyager 1 has the distinction of being the fastest object made by humans, traveling with a heliocentric velocity of 20 km/second. Like Pioneers 10 and 11, it continues to send radio signals as it travels to the edge of the solar system and joins them as part of the first exploration of deep space beyond.

Large Format Camera (LFC)
Scheduled for Shuttle launch, January 1984
Mission duration: 7 days
Objective: To acquire synoptic, high-resolution images of Earth's surface. The photographs will verify the LFC performance and will be used for making maps, interpreting geological features, and mineral exploration.

Materials Science Laboratory
Scheduled for Shuttle launch, March 1984
Mission duration: 6 days
Objective: To perform materials processing experiments in a zero-gravity space environment that allows in-flight monitoring of phenomena, sample production, and postflight analysis of samples. A series of flights of the MSL is planned at 6 to 12-month intervals.

Solar Array Flight Experiment (SAFE)
Scheduled for Shuttle launch, May 1984
Solar arrays are huge wings which convert sunlight into energy to supply power to satellites and other free-flying modules. Future arrays will be more lightweight, efficient, and flexible than those used in past programs. NASA will fly a new single-wing configuration solar array on STS-14 for a seven-day test.

Voyager 2
January 24, 1986
A flyby of Uranus at a distance of 107,000 km (66,000 mi) making measurements and taking pictures that will provide the first close-up look at the planet.

August 1989
The first flyby of Neptune.

ASTRO
Launches: March 1986
November 1986
July 1987
Mission duration: 7 days
Objective: To obtain ultraviolet (UV) data on astronomical objects, using three independent but complementary UV astronomical telescopes and a pair of auxiliary wide-field film cameras. A three-mission program is planned, the first to occur during the 1985–86 apparition of Halley's Comet.

Space Telescope (ST)
Scheduled for Shuttle launch, 1986
Above Earth's atmosphere, the ST's 2.4-m (96-in) mirror will be able to observe 350 times the volume of space now visible from ground-based observatories. Its combination of high resolution, increased sensitivity, and relatively large aperture should allow it to "see" objects 50 times fainter than can be seen from Earth.

The ST is 13 m (14.09 ft) long, 4.3 m (4.6 ft) in diameter, and weighs 11,500 kgs (25,500 lbs); it will be placed in a circular orbit 600 kms (375 mi) above Earth.

Space Telescope Science Institute (STSI)
The STSI is located at the Johns Hopkins University in Baltimore. The Institute will perform critical mission science activity for the ST, hosting astronomers who will come to the facility to use the telescope much as they would a ground-based observatory. It will process, archive, and publicize the ST's findings.

Galileo
Scheduled for Shuttle launch, 1986
A two-spacecraft mission to Jupiter: one spacecraft is an orbiter, the other a probe that will descend into Jupiter's atmosphere, measuring its chemical composition. The orbiter will photograph the planet's clouds, weather, and satellites, and measure its magnetic fields for a year or more. Objective: Information about the origin and evolution of the solar system through the Jovian system, which may reveal new insights into large-scale planetary phenomena.

Cosmic Background Explorer (COBE)
1987
Objective: To make a definitive exploration and study of the diffuse radiation of the universe between the wavelengths of 1 micrometer and 916 millimeters; this includes the cosmic background radiation thought to be the residual radiation from the Big Bang that is presumed to have started the expansion of the universe.

Tethered Satellite System
Scheduled for Shuttle launch, 1987
A data-gathering satellite to be carried into orbit by the Shuttle, then released from the cargo bay on a tether—a super-strong cord that can be 97 kms (60 mi) long—and trolled through Earth's upper atmosphere. Key elements of the system include a reel and an extendable boom, which are essential to the deployment and retrieval of the satellite.

Venus Radar Mapper (VRM)
Proposed for launch in 1988
Objective: To provide a radar map of the cloud-enveloped surface of Venus, with a resolution of one km (.6214 mi); it will carry a radar altimeter to provide topographic information and a gravity sensor.

Gamma Ray Observatory (GRO)
Proposed for launch in 1988
Newest in a long line of astronomical observatories, the GRO will observe extremely energetic particles emitted from the active nuclei of galaxies. These particles, very high energy X-rays and gamma rays, can provide information on the formation of elements and on the evolution of galaxies.

Artist's concept of the Venus Radar Mapper.

Solar Optical Telescope (SOT)
Scheduled for launch in October 1989
Mission duration: Up to 14 days
A 1-meter class multi-user facility with very high spatial resolutions, the SOT will be used to study small-scale, dynamic phenomena on the Sun's surface.

The Space Telescope, scheduled for launch into orbit by the Space Shuttle in 1986.

Space Flight

Space Transportation System (STS)
Regularly scheduled runs of the Space Shuttle will make spaceflight in the 1980s a routine occurrence. Payloads and crews for some Shuttle flights have been announced. In the following list, the crews are named in order of Commander (C), Pilot (P), Mission Specialists (MS), and Payload Specialists (PS).

STS-11, Columbia
January 1984
Mission duration: 7 days
C: Vance D. Brand
P: Robert L. Gibson
MS: Bruce McCandless II
MS: Robert L. Stewart
MS: Dr. Ronald E. McNair
Payload: Palapa B-2, Indonesian communications satellite; Large Format Camera (LFC); Payload Deployment and Retrieval System test article, a 4.9 × 4.6-m (16x15-ft) rectangular frame with lead ballast.

STS-12, Discovery
March 1984
C: Henry W. Hartsfield
P: Michael L. Coats
MS: Dr. Judith A. Resnick
MS: Dr. Stephen A. Hawley
MS: Richard M. Mullane
First launch of the third operational orbiter Discovery. Payload: Tracking and Data Relay Satellite-C (TDRS-C) and Materials Experiment Assembly (MEA).

STS-13, Challenger
April 1984
C: Robert L. Crippen
P: Francis R. Scobee
MS: Dr. George D. Nelson
MS: Terry J. Hart
MS: Dr. James D. van Hoften
Payload: Long Duration Exposure Facility (LDEF), a free-flying passive satellite for long-term experiments in space which will be retrieved on a later flight. The crew will retrieve the Solar Maximum Mission satellite, repair it, and deploy it back into space.

Artist's concept of the Long Duration Exposure Facility in the bay of the Shuttle orbiter.

STS-14, Columbia
May 1984
Payload: Communications satellites SYNCOM-IV-I, TELESAT-I, and RCA-K; Office of Aeronautics and Space Technology 1 (OAST-1) experiments package; Solar Array Flight Experiment.

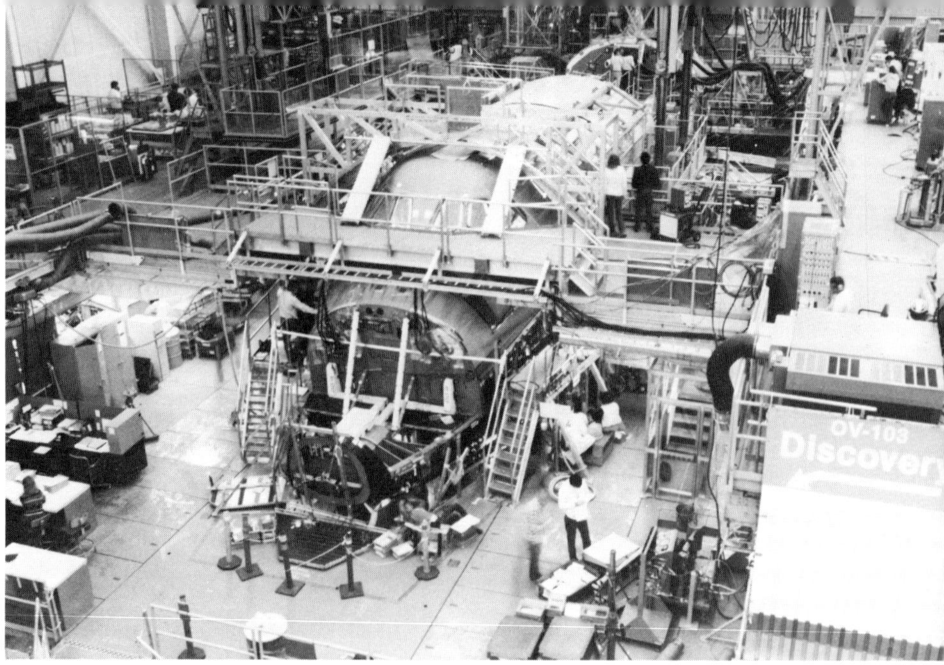

The third Shuttle orbiter, *Discovery,* in construction at the Rockwell International Palmdale (CA) facility. Delivery date: late 1983.

STS-15, Discovery
June 1984
Payload: TDRS-D, last of the Tracking and Data Relay Satellite series, and SBS-D, a commercial communications satellite.

STS-16, Challenger
July 1984
Payload: Department of Defense

STS-17, Columbia
August 1984
Payload: Earth Radiation Budget Explorer Satellite (ERBE); OSTA-3 scientific experiments; Large Format Camera (LFC); and Spartan-1, a small astronomical payload.

STS-18, Columbia
September 1984
C: Robert F. Overmyer
P: Frederick D. Gregory
MS: Dr. Don L. Lind
MS: Dr. Norman E. Thagard
MS: Dr. William E. Thornton
PS: To be named
Payload: Spacelab 3, the first operational Spacelab mission carrying experiments in materials processing, space technology, and life sciences; ARABSAT-A, an Arabian satellite; Telstar 3-C; SYNCOM-IV-2; and Westar VII.

Visions – the 1990s

Aeronautics

The future is the investigations, questions, and answers of today's scientists and engineers.

Commuter Aircraft
Nose-mounted horizontal wings and rear-mounted twin prop engines represent some of NASA's latest thinking about the shape of things to come in commuter aircraft.

Solar High Altitude Power Platform (Solar HAPP)
A Sun vehicle for the 1990s, Solar HAPP will have solar cells covering both sides of vertical stabilizers and wingtips. Daytimes the wingtips hinge up to catch maximum solar power,

Multi-body Transport.

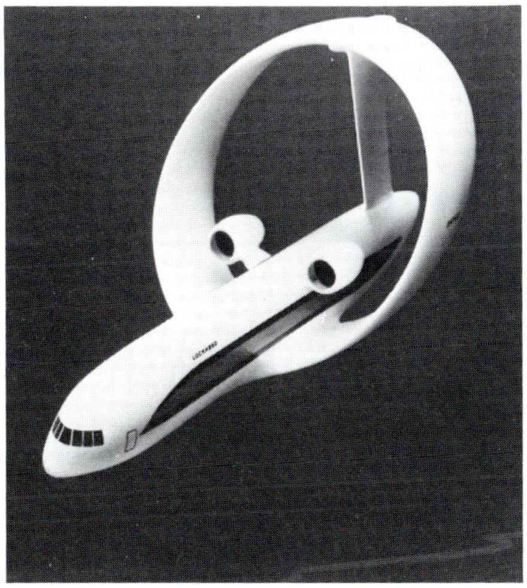

Ring Wing aircraft.

nighttimes they are horizontal for better aerodynamic performance. Launched on fuel cell power at 3 a.m. when winds are minimal, HAPP would spiral up to 21.3 km (70,000 ft) in four hours and could maintain station for months.

Multibody Transport
NASA and the Lockheed-Georgia Co. are studying a large multibody transport aircraft concept that offers potential benefits in weight, performance, cost, and fuel efficiency over conventional single-body aircraft.

Ring Wing
On the drawing board for the post-2000 year time period, Lockheed-Georgia Co. aircraft designers have a Ring Wing configuration—the wings curl upward to meet above the fuselage.

Space

In 25 years our world has left the limitations of Earth, moved outward beyond the solar system and backward toward the beginning of time. In the next quarter century new satellites and space probes, observatories and telescopes, space systems and space platforms will open other doors to the use of space, expand our knowledge of the universe, and continually take us to new frontiers.

"There can be no thought of finishing, for 'aiming at the stars,' both literally and figuratively, is a problem to occupy generations, so that no matter how much progress one makes, there is always the thrill of just beginning. . . ."

Robert H. Goddard
Letter to H. G. Wells, 1932

Space Science

Titan Probe/Radar Mapper
Proposed for launch, 1988–1992
Objective: To examine the atmosphere and surface of Saturn's largest moon; a package of instruments will parachute through the atmosphere and use the probe carrier to send back radar images of the surface. If upgraded, the mission will also employ a full-scale Saturn orbiter to study the planet, its satellites, and its rings.

Shuttle Infrared Telescope Facility (SIRTF)
1990
A 1-meter class, cryogenically-cooled, multiuser facility consisting of a telescope and associated focal plane instruments for infrared astronomy and astrophysics. SIRTF will be deployed on the Shuttle and will remain attached to it as a Spacelab payload during astronomical observations; it will be returned to Earth for refurbishment and reflight.

Mars Geoscience/Climatology Orbiter
Proposed for launch in 1990
Objective: To orbit Mars for one Martian year gathering information on the planet's surface composition, magnetic field, and season cycles of carbon dioxide, water, and dust that interact between the surface and atmosphere.

Mariner Mark II
1990s
A new generation of deep-space planetary Explorers being developed at the Jet Propulsion Laboratory. A new concept in spacecraft design, it will incorporate technological breakthroughs that will be standardized for use on many spacecraft and for a variety of missions.

Starlab
1990s
An Australia-Canada-USA orbiting 1-meter telescope to be placed on a free-flying space

Starlab.

platform for a series of six-month missions. With unique capabilities for astrophysical problems in optical and ultraviolet spectral regions, Starlab will complement the Space Telescope in imagery and spectroscopy.

Space Structures

The Shuttle's ability to carry cargoes into space has led to designs for complicated space structures—giant platforms and antennas that will take us to the next century, to a time when space stations may become the first space settlements.

Space Platforms
1988–1990
Space platforms, large multi-purpose systems, will have a central "bus" to house the power generator and electronic and thermal systems; modules will carry communications equipment, scientific experiment payloads, a telescope, a communications dish. The basic Space Platform can evolve into advanced systems, including, with the addition of a life-supporting module, a Science and Applications Manned Space Platform.

Large Space Antennas
1990s
Large space antennas that will fold into containers on Earth, will go up whole in one Shuttle trip and deploy automatically in space. They will revolutionize worldwide communications—a few in high geostationary orbits will cover the globe while millions of small home rooftop dishes will receive the signals. They also will have applications for remote sensing of Earth resources and spacecraft tracking.

Several designs have been explored—hoop-column (or maypole) antenna, offset wrap-rib antenna, wire-wheel antenna—all measuring in the 100-meter (328-ft) class.

Then will come building the structures in space. An Automated Beam Builder has been designed, a machine that will sit at the end of the Shuttle's cargo bay and heat, shape, and weld material into meter-wide triangular beams. The beams will be cut to desired lengths and joined together to create large structures of any configuration. The early large space platforms and antennas could become the first space settlements.

The next twenty-five years. Eagle Engineering concept of lunar mining operations to produce liquid oxygen from ilmenite, an oxygen rich component of lunar soil. The lunar lander at the top background of the picture would transport the liquid oxygen to an orbiting space station.

For the Classroom

1. Ask your students to list what they think will be included in NASA's 50th anniversary book.
2. Research topics:
   - Life in the year 2008
   - Space stations
   - Transportation of the future
   - Solar power stations
3. Discuss new uses for the Shuttle during the next two decades.
4. Discuss what astronomers might learn with the advanced telescopes of the 1980s and 90s.
5. Have your students read a current science fiction novel or see a science fiction film that involves large space structures and write a critique of the hardware created for the story.
5. Keep a clipping file on the Pioneer and Voyager journeys beyond the solar system.

Appendix I

NASA Major Launch Record, 1958–1983

This list is a compilation of launches by NASA of (1) payloads that went into orbit or that achieved an altitude of at least 6400 kilometers (3978 miles) and (2) major suborbital flight tests or experiments (noted with an asterisk beside the name). Dates given are determined by local time at the launch site. Not listed are launch failures or sounding rocket launches.

Name	Launch Date	Launch Vehicle*
Pioneer 1	11 October 1958	Thor-Able I
Pioneer 3	6 December 1958	Juno II
Vanguard 2	17 February 1959	Vanguard
Pioneer 4	3 March 1959	Juno II
Explorer 6	7 August 1959	Thor-Able III
Big Joe*	9 September 1959	Atlas
Vanguard 3	18 September 1959	Vanguard
Little Joe 1*	4 October 1959	Little Joe
Explorer 7	13 October 1959	Juno II
Shotput 1*	28 October 1959	Shotput
Little Joe 2*	4 November 1959	Little Joe
Little Joe 3*	4 December 1959	Little Joe
Shotput 2*	16 January 1960	Shotput
Little Joe 4*	21 January 1960	Little Joe
Shotput 3*	27 February 1960	Shotput
Pioneer 5	11 March 1960	Thor-Able IV
Shotput 4*	1 April 1960	Shotput
TIROS 1	1 April 1960	Thor-Able
Shotput 5*	31 May 1960	Shotput
Echo 1	12 August 1960	Thor-Delta
Scout 2*	4 October 1960	Scout
Explorer 8	3 November 1960	Juno II
Little Joe 5*	8 November 1960	Little Joe
TIROS 2	23 November 1960	Thor-Delta
Mercury-Redstone 1A*	19 December 1960	Redstone
Mercury-Redstone 2*	31 January 1961	Redstone
Explorer 9	16 February 1961	Scout
Mercury-Atlas 2*	21 February 1961	Atlas
Little Joe 5A*	18 March 1961	Little Joe
Mercury-Redstone BD*	24 March 1961	Redstone
Explorer 10	25 March 1961	Thor-Delta
Explorer 11	27 April 1961	Juno II
Little Joe 5B*	28 April 1961	Little Joe
Freedom 7* (Mercury-Redstone 3)	5 May 1961	Redstone
TIROS 3	12 July 1961	Thor-Delta
Liberty Bell 7* (Mercury-Redstone 4)	21 July 1961	Redstone
Explorer 12	16 August 1961	Thor-Delta
Ranger 1	23 August 1961	Atlas-Agena B
Explorer 13	25 August 1961	Scout
Mercury-Atlas 4	13 September 1961	Atlas D
Probe A (P-21)*	19 October 1961	Scout
Saturn-Apollo 1* (SA-1)	27 October 1961	Saturn I
Ranger 2	18 November 1961	Atlas-Agena B
Mercury-Atlas 5	29 November 1961	Atlas D
Echo* (AVT-1)	15 January 1962	Thor
Ranger 3	26 January 1962	Atlas-Agena B
TIROS 4	8 February 1962	Thor-Delta
Friendship 7 (Mercury-Atlas 6)	20 February 1962	Mercury-Atlas D
Reentry 1*	1 March 1962	Scout
OSO-1	7 March 1962	Thor-Delta
Probe B (P-21A)*	29 March 1962	Scout
Ranger 4	23 April 1962	Atlas-Agena B
Saturn-Apollo 2* (SA-2)	25 April 1962	Saturn I
Ariel 1	26 April 1962	Thor-Delta
Aurora 7	24 May 1962	Atlas D

*Thor-Delta launch vehicle configurations are abbreviated as follows: Thor-Delta (Thor-Delta, Thor-improved Delta), TAT-Delta (thrust-augmented Thor-Delta), TAID (thrust-augmented Thor-improved Delta), LTTAT-Delta (long-tank, thrust-augmented Thor-improved Delta), TAT-Agena (thrust-augmented Thor-Agena).

Name	Launch Date	Launch Vehicle*
(Mercury-Atlas 7)		
TIROS 5	19 June 1962	Thor-Delta
Telstar 1	10 July 1962	Thor-Delta
Echo (AVT-2)*	18 July 1962	Thor
Mariner 2	27 August 1962	Atlas-Agena B
TIROS 6	18 September 1962	Thor-Delta
Alouette 1	28 September 1962	Thor-Agena B
Explorer 14	2 October 1962	Thor-Delta
Sigma 7 (Mercury-Atlas 8)	3 October 1962	Atlas D
Ranger 5	18 October 1962	Atlas-Agena B
Explorer 15	27 October 1962	Thor-Delta
Saturn-Apollo 3* (SA-3)	16 November 1962	Saturn I
Relay 1	13 December 1962	Thor-Delta
Explorer 16	16 December 1962	Scout
Syncom 1	14 February 1963	Thor-Delta
Saturn-Apollo 4* (SA-4)	28 March 1963	Saturn I
Explorer 17	2 April 1963	Thor-Delta
Telstar 2	7 May 1963	Thor-Delta
Faith 7 (Mercury-Atlas 9)	15 May 1963	Atlas D
TIROS 7	19 June 1963	Thor-Delta
Syncom 2	26 July 1963	Thor-Delta
Little Joe II*	28 August 1963	Little Joe II
Explorer 18	26 November 1963	Thor-Delta
Atlas-Centaur 2	27 November 1963	Atlas-Centaur
Explorer 19	19 December 1963	Scout
TIROS 8	21 December 1963	Thor-Delta
Relay 2	21 January 1964	Thor-Delta
Echo 2	25 January 1964	Thor-Agena B
Saturn-Apollo 5 (SA-5)	29 January 1964	Saturn I
Ranger 6	30 January 1964	Atlas-Agena B
Ariel 2	27 March 1964	Scout
Gemini-Titan I	8 April 1964	Gemini-Titan II
FIRE 1*	14 April 1964	Atlas D
Apollo (A-001)* (Transonic Abort)	13 May 1964	Little Joe II
Apollo-Saturn 101 (SA-6)	28 May 1964	Saturn I
Atlas-Centaur 3*	30 June 1964	Atlas-Centaur
SERT 1*	20 July 1964	Scout
Ranger 7	28 July 1964	Atlas-Agena B
Reentry 4*	18 August 1964	Scout
Syncom 3	19 August 1964	TAT-Delta
Explorer 20	25 August 1964	Scout
Nimbus 1	28 August 1964	Thor-Agena B
OGO-1	5 September 1964	Atlas-Agena B
Apollo-Saturn 102 (SA-7)	18 September 1964	Saturn I
Explorer 21	4 October 1964	Thor-Delta
Explorer 22	9 October 1964	Scout
Mariner 3	5 November 1964	Atlas-Agena D
Explorer 23	6 November 1964	Scout
Explorer 24 and Explorer 25 (Injun 4)	21 November 1964	Scout
Mariner 4	28 November 1964	Atlas-Agena D
Apollo Maximum Q Abort*	8 December 1964	Little Joe II
Atlas-Centaur 4	11 December 1964	Atlas-Centaur
San Marco 1	15 December 1964	Scout
Explorer 26	21 December 1964	Thor-Delta
Gemini-Titan 2*	19 January 1965	Titan II
TIROS 9	22 January 1965	Thor-Delta
OSO-2	3 February 1965	Thor-Delta
Pegasus 1 and Apollo-Saturn 103 (SA-9)	16 February 1965	Saturn I
Ranger 8	17 February 1965	Atlas-Agena B
Ranger 9	21 March 1965	Atlas-Agena B
Gemini 3	23 March 1965	Titan II
Early Bird I	6 April 1965	TAT-Delta
Explorer 27	29 April 1965	Scout
FIRE 2*	22 May 1965	Atlas D
Pegasus 2 and Apollo-Saturn 104 (SA-8)	25 May 1965	Saturn I
Explorer 28	29 May 1965	Thor-Delta
Gemini 4	3 June 1965	Titan II
TIROS 10	2 July 1965	TAT-Delta
Pegasus 3 and Apollo-Saturn 105 (SA-10)	30 July 1965	Saturn I
Scout (SEV-A)	10 August 1965	Scout
Centaur-Surveyor	11 August 1965	Atlas-Centaur
Gemini 5	21 August 1965	Titan II
OGO-2	14 October 1965	TAT-Agena D
Explorer 29 (GEOS-1)	6 November 1965	TAID
Explorer 30	19 November 1965	Scout
Alouette 2 and Explorer 31	28 November 1965	Thor-Agena B
Gemini 7	4 December 1965	Titan II
FR-1	6 December 1965	Scout
Gemini 6	15 December	Titan II

Name	Launch Date	Launch Vehicle*
	1965	
Pioneer 6	16 December 1965	TAID
Apollo (A-004)* Intermediate Altitude Abort	20 January 1966	Little Joe II
ESSA 1	3 February 1966	Thor-Delta
Reentry 5*	9 February 1966	Scout
Apollo-Saturn 201*	26 February 1966	Saturn IB
ESSA 2	28 February 1966	TAID
Gemini-Agena Target Vehicle 8	16 March 1966	Atlas-Agena D
Gemini 8	16 March 1966	Titan II
Centaur-Surveyor	7 April 1966	Atlas-Centaur
OAO-1	8 April 1966	Atlas-Agena D
Nimbus 2	15 May 1966	TAT-Agena B
Explorer 32	25 May 1966	Thor-Delta
Surveyor 1	30 May 1966	Atlas-Centaur
Augmented Target Docking Adapter	1 June 1966	Atlas D
Gemini 9	3 June 1966	Titan II
OGO-3	7 June 1966	Atlas-Agena B
PAGEOS 1	23 June 1966	TAT-Agena D
Explorer 33	1 July 1966	TAID
Apollo-Saturn 203	5 July 1966	Saturn IB
Gemini-Agena Target Vehicle 10	18 July 1966	Atlas-Agena D
Gemini 10	18 July 1966	Titan II
Lunar Orbiter 1	10 August 1966	Atlas-Agena D
Pioneer 7	17 August 1966	TAID
Apollo-Saturn 202*	25 August 1966	Saturn IB
Gemini-Agena Target Vehicle 11	12 September 1966	Atlas-Agena D
Gemini 11	12 September 1966	Titan II
Surveyor 2	20 September 1966	Atlas-Centaur
ESSA 3	2 October 1966	TAID
Centaur-Surveyor	26 October 1966	Atlas-Centaur
Intelsat-II F-1	27 October 1966	TAID
Lunar Orbiter 2	6 November 1966	Atlas-Agena D
Gemini-Agena Target Vehicle 12	11 November 1966	Atlas-Agena D
Gemini 12	11 November 1966	Titan II
ATS-1	6 December 1966	Atlas-Agena D
Biosatellite 1	14 December 1966	TAID
Intelsat-II F-2	11 January 1967	TAID
ESSA 4	26 January 1967	TAID
Lunar Orbiter 3	5 February 1967	Atlas-Agena D
OSO-3	8 March 1967	Thor-Delta
Intelsat-II F-3	23 March 1967	TAID
ATS-2	6 April 1967	Atlas-Agena D
Surveyor 3	17 April 1967	Atlas-Centaur
ESSA 5	20 April 1967	TAID
San Marco 2	26 April 1967	Scout
Lunar Orbiter 4	4 May 1967	Atlas-Agena D
Ariel 3	5 May 1967	Scout
Explorer 34	24 May 1967	TAID
Mariner 5	14 June 1967	Atlas-Agena D
Surveyor 4	14 July 1967	Atlas-Centaur
Explorer 35	19 July 1967	TAID
OGO-4	28 July 1967	TAT-Agena D
Lunar Orbiter 5	1 August 1967	Atlas-Agena D
Biosatellite 2	7 September 1967	TAID
Surveyor 5	8 September 1967	Atlas-Centaur
Intelsat-II F-4	28 September 1967	TAID
OSO-4	18 October 1967	Thor-Delta
ATS-3	5 November 1967	Atlas-Agena D
Surveyor 6	7 November 1967	Altas-Centaur
Apollo 4	9 November 1967	Saturn V
ESSA 6	10 November 1967	TAID
Pioneer 8 and TTS 1	13 December 1967	TAID
Surveyor 7	7 January 1968	Atlas-Centaur
Explorer 36 (GEOS-2)	11 January 1968	TAID
Apollo 5 (Apollo-Saturn 204)	22 January 1968	Saturn IB
OGO-5	4 March 1968	Atlas-Agenda D
Explorer 37	5 March 1968	Scout
Apollo 6	4 April 1968	Saturn V
Reentry 6*	27 April 1968	Scout
IRIS-1	16 May 1968	Scout
Explorer 38	4 July 1968	TAID
Explorers 39 and 40	8 August 1968	Scout
ATS-4	10 August 1968	Atlas-Centaur
ESSA 7	16 August 1968	LTTAT-Delta
Aurorae	3 October 1968	Scout
Apollo 7	11 October 1968	Saturn IB
Pioneer 9 and TETR 2	8 November 1968	TAID
HEOS-1	5 December 1968	TAID
OAO-2	7 December 1968	Atlas-Centaur
ESSA 8	15 December 1968	LTTAT-Delta
Intelsat-III F-2	18 December 1968	LTTAT-Delta
Apollo 8	21 December 1968	Saturn V
OSO-5	22 January 1969	Thor-Delta
ISIS-1	30 January 1969	TAID

Name	Launch Date	Launch Vehicle*
Intelsat-III F-3	5 February 1969	LTTAT-Delta
Mariner 6	25 February 1969	Atlas-Centaur
ESSA 9	26 February 1969	TAID
Apollo 9	3 March 1969	Saturn V
Mariner 7	27 March 1969	Atlas-Centaur
Nimbus 3	14 April 1969	LTTAT-Agena D
Apollo 10	18 May 1969	Saturn V
Intelsat-III F-4	22 May 1969	LTTAT-Delta
OGO-6	5 June 1969	LTTAT-Agena D
Explorer 41	21 June 1969	TAID
Biosatellite 3	29 June 1969	LTTAT-Delta
Apollo 11	16 July 1969	Saturn V
Intelsat-III F-5	26 July 1969	LTTAT-Delta
OSO-6 and PAC	9 August 1969	LTTAT-Delta
ATS-5	12 August 1969	Atlas-Centaur
Boreas	1 October 1969	Scout
Azur	8 November 1969	Scout
Apollo 12	14 November 1969	Saturn V
Skynet 1	22 November 1969	LTTAT-Delta
Intelsat-III F-6	14 January 1970	LTTAT-Delta
ITOS-1 and OSCAR 5	23 January 1970	LTTAT-Delta
SERT 2	4 February 1970	LTTAT-Agena D
NATOSAT 1	20 March 1970	LTTAT-Delta
Nimbus 4	8 April 1970	LTTAT-Agena D
Apollo 13	11 April 1970	Saturn V
Intelsat-III F-7	22 April 1970	LTTAT-Delta
Intelsat-III F-8	23 July 1970	LTTAT-Delta
OFO and RMS	9 November 1970	Scout
NOAA 1 and CEPE	11 December 1970	LTTAT-Delta
Explorer 42	12 December 1970	Scout
Intelsat-IV F-2	25 January 1971	Atlas-Centaur
Apollo 14	31 January 1971	Saturn V
NATOSAT 2	2 February 1971	LTTAT-Delta
Explorer 43	13 March 1971	TAID
ISIS-2	31 March 1971	TAID
San Marco 3	24 April 1971	Scout
Mariner 9	30 May 1971	Atlas-Centaur
PAET	20 June 1971	Scout
Explorer 44	8 July 1971	Scout
Apollo 15	26 July 1971	Saturn V
Eole	16 August 1971	Scout
OSO-7	29 September 1971	LTTAT-Delta
Explorer 45	15 November 1971	Scout
Ariel 4	11 December 1971	Scout
Intelsat-IV F-3	19 December 1971	Atlas-Centaur
Intelsat-IV F-4	22 January 1972	Atlas-Centaur
HEOS-2	31 January 1972	LTTAT-Delta
Pioneer 10	3 March 1972	Atlas-Centaur-TE-M-364-4
TD-1A	12 March 1972	LTTAT-Delta
Apollo 16	16 April 1972	Saturn V
Intelsat-IV F-5	13 June 1972	Atlas-Centaur
Landsat 1 (ERTS-1)	23 July 1972	LTTAT-Delta
Explorer 46	13 August 1972	Scout
OAO-3	21 August 1972	Atlas-Centaur
Explorer 47	22 September 1972	LTTAT-Delta
NOAA 2 and OSCAR 6	15 October 1972	LTTAT-Delta
Anik 1	9 November 1972	LTTAT-Delta
Explorer 48	16 November 1972	Scout
ESRO 4	21 November 1972	Scout
Apollo 17	7 December 1972	Saturn V
Nimbus 5	11 December 1972	LTTAT-Delta
Aeros	16 December 1972	Scout
Pioneer 11	6 April 1973	Atlas-Centaur-TE-M-364-4
Anik 2	20 April 1973	LTTAT-Delta
Skylab 1	14 May 1973	Saturn V
Skylab 2	25 May 1973	Saturn IB
Explorer 49	10 June 1973	LTTAT-Delta
Skylab 3	28 July 1973	Saturn IB
Intelsat-IV F-7	23 August 1973	Atlas-Centaur
Explorer 50	25 October 1973	LTTAT-Delta
Mariner 10	3 November 1973	Atlas-Centaur
NOAA 3	6 November 1973	LTTAT-Delta
Skylab 4	16 November 1973	Saturn IB
Explorer 51	16 December 1973	LTTAT-Delta
Skynet IIA	18 January 1974	LTTAT-Delta
San Marco 4	18 February 1974	Scout
UK-X4	8 March 1974	Scout
Westar 1	13 April 1974	LTTAT-Delta
SMS-1	17 May 1974	LTTAT-Delta
ATS-6	30 May 1974	Titan IIIC
Explorer 52	3 June 1974	Scout
Aeros 2	16 July 1974	Scout
ANS	30 August 1974	Scout
Westar 2	10 October 1974	LTTAT-Delta
Ariel 5 (UK-5)	15 October 1974	Scout
NOAA 4 and OSCAR 7 and INTASAT	15 November 1974	LTTAT-Delta
Intelsat-IV F-8	21 November 1974	Atlas-Centaur
Skynet IIB	22 November 1974	LTTAT-Delta
Helios 1	10 December 1974	Titan IIIE-Centaur-TE-M-364-4
Symphonie A	17 December 1974	LTTAT-Delta
Landsat 2	22 January 1975	Delta

Name	Launch Date	Launch Vehicle*
SMS-2	6 February 1975	Delta
Intelsat-IV F-6	20 February 1975	Atlas-Centaur
GEOS-3	9 April 1975	Delta
Explorer 53	7 May 1975	Scout
Anik 3	7 May 1975	Delta
Intelsat-IV F-1	22 May 1975	Atlas-Centaur
Nimbus 6	12 June 1975	Delta
OSO-1	21 June 1975	Delta
Apollo-Soyuz Test Project	15 July 1975	Saturn 1B
COS-B	8 August 1975	Delta
Viking 1	20 August 1975	Titan III Centaur
Symphonie B	26 August 1975	Delta
Viking 2	9 September 1975	Titan III Centaur
Intelsat-IVA F-1	25 September 1975	Atlas-Centaur
Explorer 54	6 October 1975	Delta
GOES/NOAA	16 October 1975	Delta
Explorer 55	20 November 1975	Delta
RCA-A	13 December 1975	Delta
Helios 2	15 January 1976	Titan III Centaur
CTS	17 January 1976	Delta
Intelsat-IVA F-2	29 January 1976	Atlas-Centaur
Marisat-A	19 February 1976	Delta
RCA-B	26 March 1976	Delta
NATO-III A	22 April 1976	Delta
LAGEOS	4 May 1976	Delta
Comstar	13 May 1976	Atlas-Centaur
Marisat-B	9 June 1976	Delta
Gravity Probe A*	18 June 1976	Scout
Palapa A1	8 July 1976	Delta
Comstar 2	22 July 1976	Atlas-Centaur
ITOS-H/NOAA5	29 July 1976	Delta
Marisat-C	14 October 1976	Delta
NATO-3 B	27 January 1977	Delta
Palapa A2	10 March 1977	Delta
GEOS/ESA	20 April 1977	Delta
Intelsat-IVA F-4	26 May 1977	Atlas-Centaur
GOES/NOAA	16 June 1977	Delta
GMS-Japan	14 July 1977	Delta
HEAO-1	12 August 1977	Atlas-Centaur
Voyager 2	20 August 1977	Titan III Centaur
SIRIO-I	25 August 1977	Delta
Voyager 1	5 September 1977	Titan III Centaur
ISEE-1 and 2	22 October 1977	Delta
Meteosat/ESA	22 November 1977	Delta
CS/Japan	14 December 1977	Delta
Intelsat-IVA F-3	7 January 1978	Atlas-Centaur
IUE	26 January 1978	Delta
Landsat 3	5 March 1978	Delta
Intelsat-IVA F-6	31 March 1978	Atlas-Centaur
BSE/Japan	7 April 1978	Delta
HCMM	26 April 1978	Scout
OTS/ESA	11 May 1978	Delta
Pioneer Venus 1	20 May 1978	Atlas-Centaur
GOES-3/NOAA	16 June 1978	Delta
Seasat	26 June 1978	Atlas-F
Comstar 3	29 June 1978	Atlas-Centaur
GEOSAB/ESA	14 July 1978	Delta
Pioneer Venus 2	8 August 1978	Atlas-Centaur
ISEE-3	12 August 1978	Delta
TIROS-N	13 October 1978	Atlas-F
Nimbus 7	24 October 1978	Delta
HEAO-2	13 November 1978	Atlas-Centaur
NATO-III-C	18 November 1978	Delta
Anik 4	15 December 1978	Delta
SAGE	18 February 1979	Scout
UK-6 (Ariel)	2 June 1979	Scout
NOAA 6	27 June 1979	Atlas-E/F
Westar 3	9 August 1979	Delta
HEAO-3	20 September 1979	Atlas-Centaur
Magsat	30 October 1979	Scout
SATCOM/RCA	6 December 1979	Delta
SMM	14 February 1980	Delta
NOAA 7	29 May 1980	Atlas-F
GOES-D	9 September 1980	Delta
SBS-A	15 November 1980	Delta
Intelsat V-A	6 December 1980	Atlas-Centaur
Comstar 4	21 February 1981	Atlas-Centaur
Space Shuttle Columbia	12 April 1981	STS-1
GOES 5	22 May 1981	Delta
Intelsat V-B	23 May 1981	Atlas-Centaur
NOAA C	23 June 1981	Atlas-F
Dynamics Explorer	3 August 1981	Delta
SBS-B	24 September 1981	Delta
SME	6 October 1981	Delta
Space Shuttle Columbia	12 November 1981	STS-2
RCA-D	19 November 1981	Delta
Intelsat V-C	15 December 1981	Atlas-Centaur
RCA-C	15 January 1982	Delta
Westar 4	25 February 1982	Delta
Intelsat V-D	4 March 1982	Atlas-Centaur
Space Shuttle Columbia	22 March 1982	STS-3
INSAT 1A	10 April 1982	Delta
Westar 5	8 June 1982	Delta
Space Shuttle Columbia	27 June 1982	STS-4
Landsat 4	16 July 1982	Delta

Name	Launch Date	Launch Vehicle*
Anik D1	26 August 1982	Delta
Intelsat V-E	28 September 1982	Atlas-Centaur
RCA-E	27 October 1982	Delta
Space Shuttle Columbia	11 November 1982	STS-5
SBS-C	11 November 1982	STS-5
Anik C	12 November 1982	STS-5
IRAS	25 January 1983	Delta
NOAA 8	28 March 1983	Atlas-F
Space Shuttle Challenger	4 April 1983	STS-6
TDRS-1	4 April 1983	STS-6
RCA-F	11 April 1983	Delta
GOES-F	28 April 1983	Delta
Intelsat V-F	19 May 1983	Atlas-Centaur
EXOSAT	26 May 1983	Delta
Space Shuttle Challenger	June 1983	STS-7
Anik C	June 1983	STS-7
Palapa-B	June 1983	STS-7
Galaxy-A	June 1983	Delta
Telstar 3A	July 1983	Delta
Space Shuttle Challenger	August 1983	STS-8
TDRS-B	August 1983	STS-8
INSAT 1B	August 1983	STS-8
NOAA F	August 1983	Atlas-F
RCA-G	August 1983	Delta
Galaxy-B	September 1983	Delta
Space Shuttle Columbia	September 1983	STS-9
San Marco D/L	November 1983	Scout
Intelsat VA-A	December 1983	Atlas-Centaur

☆U.S. GOVERNMENT PRINTING OFFICE: 1984—433-521

Appendix II

NASA Educational Services

NASA's educational programs serve the teacher, the student, the school, and the community. Information about the programs and services, including lists of NASA Publications and NASA Films, is available from the Education Services Office at the NASA Centers that serve specific geographic areas. Special resource centers, which make professional curriculum materials available to teachers, have been established at most of the NASA centers. Where to write for services:

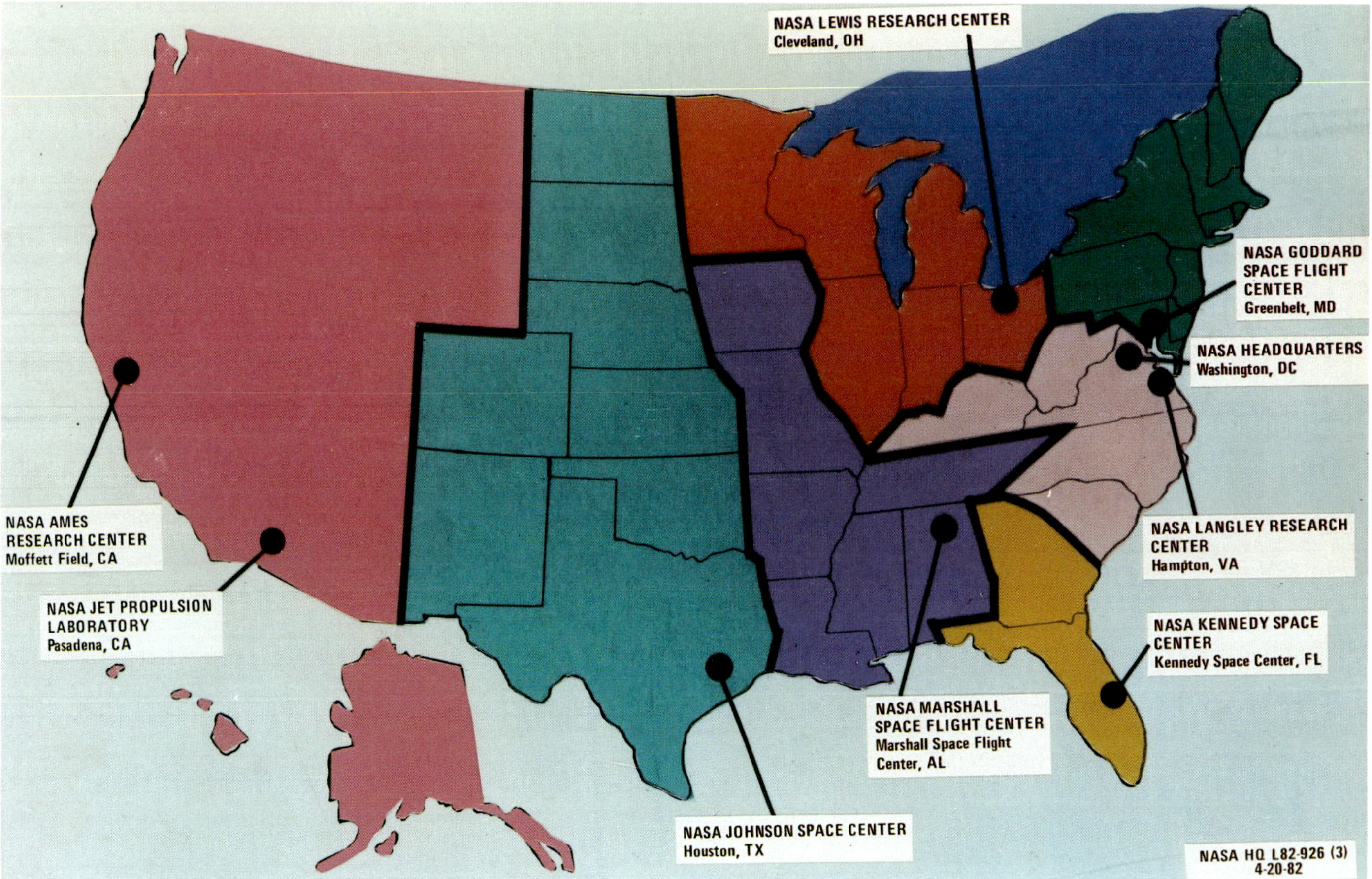

NASA Ames Research Center
Moffett Field, CA 94035
Alaska, Arizona, California, Hawaii, Idaho, Montana, Nevada, Oregon, Utah, Washington, Wyoming

NASA Goddard Space Flight Center
Greenbelt, MD 20771
Connecticut, Delaware, District of Columbia, Maine, Maryland, Massachusetts, New Hampshire, New Jersey, New York, Pennsylvania, Rhode Island, Vermont

NASA Lyndon B. Johnson Space Center
Houston, TX 77058
Colorado, Kansas, Nebraska, New Mexico, N. Dakota, Oklahoma, S. Dakota, Texas

NASA John F. Kennedy Space Center
Kennedy Space Center, FL 32899
Florida, Georgia, Puerto Rico, Virgin Islands

NASA Langley Research Center
Hampton, VA 23665
Kentucky, N. Carolina, S. Carolina, Virginia, West Virginia

NASA Lewis Research Center
Cleveland, OH 44135
Illinois, Indiana, Michigan, Minnesota, Ohio, Wisconsin

NASA George C. Marshall Space Flight Center
Marshall Space Flight Center, AL 35812
Alabama, Arkansas, Iowa, Louisiana, Mississippi, Missouri, Tennessee